Soil Water and Ground Water Sampling

Neal Wilson

About the cover: The photograph on the cover is of individual sand grains of Junction Creek Sandstone from Colorado, originally magnified approximately 25 times. In an aquifer comprised of sandstone or unconsolidated sand, ground water and contaminants flow horizontally and vertically between the sand grains.

Library of Congress Cataloging-in-Publicati

Wilson, Neal
Soil water and ground water sampling / by Neal Wilson
p. cm.
Includes bibliographical references and index.
ISBN 1-56670-073-6
1. Groundwater — Sampling. 2. Soil moisture — Sampling I. Title.
GB1001.72.S3W55 1995
553.7´9´0287—dc20 94-38733
CIP

Lewis Publishers is an imprint of CRC Press

International Standard Book Number 1-56670-073-6
Library of Congress Card Number 94-38733
Printed in the United States of America 1 2 3 4 5 6 7 8 9 0
Printed on acid-free paper.

Note: This book has been written by Neal Wilson and does not necessarily represent the views, policies, or programs of the State of Minnesota.

Preface

Ground water sampling is necessary to adequately characterize important ground water resources. If improper sampling equipment or techniques are used to obtain ground water samples, then subsequent decisions based on the sampling effort relating to human health and environmental impacts and potentially involving large sums of money will not be accurate.

The primary goal for ground water sampling is to obtain representative ground water samples and field parameters. The techniques, equipment, personnel, and documentation of the sampling effort may have to stand up to hostile scrutiny in a court of law, and this should be considered when designing and executing a sampling plan.

This text presents the interrelated components necessary to produce representative ground water samples from initial site characterization through laboratory analysis. The text is written to provide a broad context to evaluate ground water sampling programs and to enable individuals to put together, critique, or follow ground water sampling plans.

Focused research on ground water sampling over the last decade has resulted in an ability to produce ground water sampling data commensurate with its intended use. Some sampling efforts are not, however, currently producing representative ground water data due to lack of information on optimizing sampling techniques and equipment, or due to a lack of accountability.

A central tenet of environmental programs should be that at every step in the process procedures should be identified and implemented to reduce the potential for error. Data quality objectives need to be formally specified, and the best available techniques, personnel, and equipment should be provided to ensure defensible data. As more research into ground water sampling is undertaken, the techniques and equipment used for monitoring ground water may change substantially. Ground water sampling and analysis need to keep pace with research and technology to reduce costs and increase data quality.

Currently there are no accepted educational or vocational standards for those who take ground water samples. It should become apparent after reading this text that data quality objectives should require that individuals who sample ground water have periodic, formalized, up-to-date training in ground water sampling. This is to ensure that at least a minimum level of proficiency is met.

Preface *(continued)*

In addition to keeping current with the latest developments in ground water sampling techniques and equipment for subsequent editions of this book, the author is working toward standardized training/ certification for those who acquire ground water samples. Questions or comments about this book or training and certification may be sent to the author, in care of CRC Press, Inc.

The Author

Neal Wilson has a Bachelors degree in Geology from the University of Minnesota and is currently employed by the Minnesota Pollution Control Agency. Mr. Wilson has written the first guide in the nation for the design and operation of land treatment of landfill leachate systems. Some of the ground water projects he has worked on include those associated with: non-point source (agricultural) pollution; industrial and municipal waste water facilities; mixed municipal, industrial, and demolition landfills; and hazardous waste sites.

Mr. Wilson provides instruction on a contractual basis in the regulatory aspects of environmental work, specializing in "hands on" training for environmental sampling. He also audits sampling efforts of consulting firms and sampling companies.

Acknowledgments

The author would like to acknowledge the assistance of Dr. Satish Gupta (University of Minnesota Soil Science Department), Kurt Schroeder (Minnesota Pollution Control Agency), Beverly L. Herzog (Illinois State Geologic Survey), and Dr. Michael Barcelona for their review of this book. In addition, the following companies assisted in the completion of this book:

- Beckman Instruments, Inc.
  Fullerton, California 92634-3100

- Engineered Systems and Designs
  119A Sandy Drive
  Neward, Delaware 19713

- Geoguard, Inc.
  536 Orient Street
  P.O. Box 149
  Medina, New York 14103-0149

- Hach Company
  P.O. Box 389
  Loveland, Colorado 80539

- Instrumentation Northwest
  14972 N.E. 31st Circle
  Redmond, Washington 98052

- Isco Incorporated
  531 Westgate Boulevard
  Lincoln, Nebraska 68528-1486

- Keck Instruments, Inc.
  1099 W. Grand River
  P.O. Box 345
  Williamston, Michigan 48895

- Solomat, Inc.
  The Waterside Building
  26 Pearl Street
  Norwalk, Connecticut 06850

Contents

chapter one

Ground water monitoring and site remediation

Remember when discoursing about water
to adduce first experience and then reason.
Leonardo da Vinci, 1452–1519

Introduction to ground water monitoring

Ground water monitoring may be initiated when a specific land-use activity has been identified that has or could potentially negatively alter (contaminate) the natural state of the soil and/or the ground water. If contamination has occurred, the site must be adequately characterized so that decisions can be made on how to best manage the site. Monitoring is an effort to obtain an understanding of the chemical, physical, and biologic characteristics of water through statistical sampling.[11] Ground water monitoring program objectives for sites that are or could become contaminated include

- Protecting ground water users, and facility owners and operators
- Assessing facility design and operation
- Guiding regulatory and enforcement actions
- Developing and implementing remedial actions for contaminated sites
- Evaluating the effectiveness of environmental programs

Some of the more common ground water contamination problems include those associated with landfills, leaking underground storage tanks, spill sites, hazardous waste sites, septic tank systems, leaking industrial and municipal wastewater lagoons, improper management of industrial and municipal wastewater and sludge, and non-point source pollution derived from agricultural practices. A contaminant may be hazardous (ignitable, corrosive, reactive, toxic, lethal, or toxic) and cause acute or chronic (including carcinogenic) impacts to human health. Environmental damage may also result from ground water contamination (Figure 1.1).

Figure 1.1 A hazardous waste site.

Investigative plans

Once a contamination problem has been identified a plan needs to be formulated and implemented to assess the immediate risks to human health and the environment, quantify the constituents of concern, determine the horizontal and vertical rates of migration of the contaminant plume, develop ways to reduce or eliminate the contamination, and further identify and protect those potentially affected. A statement of what problem is to be solved, the information required, what the project objectives are and how the objectives are to be met must be concisely stated and sufficiently detailed to permit a clear understanding by all the parties involved in the data collection effort.[10]

Quality assurance/quality control (QA/QC) measures must be employed at every step, from monitoring point design to sample collection and analysis. Data quality objectives[9] must be taken into account when a monitoring system is being designed as must statistical considerations.[6] Data quality objectives are statements that provide definitions of confidence, which are required for interpretation of the data.[3] QA/QC and data quality objectives are presented in Chapter 7.

A phased approach is usually undertaken to adequately characterize a site. A preliminary (Phase I) site characterization consists of a records review, a site reconnaissance, and interviews with owners, occupants, and government officials.[1] The Phase I report is then generated from the compiled information.

Site use history includes title searches, insurance atlases, aerial photographs, site plans, site records, utility company records, and interviews with

Figure 1.2 Workers in level B protective suits with self-contained breathing apparatus at a hazardous waste spill site.

individuals who have lived near or worked at the site for an extended period of time.[4] Existing hydrogeologic information includes soil surveys, geologic and hydrologic maps and reports, topographic maps, and well records.

The initial site reconnaissance may include the use of an organic vapor monitor, an explosivity meter, video camera recorders, still cameras, and a hand-held soil sampler. Safety is of paramount importance — a site safety plan should be formulated, documented, approved, and adhered to for each phase of the investigation (see Chapter 6) (Figure 1.2).

Field notes for site reconnaissance should include dates, times, personnel, maps of the salient site features, and descriptions of what was found at the site. Field notes may become evidence in court, so the notes should be made in ink in a notebook with a sewn binding and be descriptive and accurate. If a video camera recorder is used, a verbal narrative should be included describing date, time, site, name of persons at the site, the direction the camera is facing (e.g., "now facing north"), and what the camera is recording. Photographs should be labeled with date and time, site, photographer, and a description of what the photograph represents. Sometimes several years may elapse before a photograph may become a critical piece of evidence; photographs should be labeled promptly after processing, so labeling is not neglected.

Existing site information may be expanded with direct, field-investigation methods including borings, test pits, piezometers, monitoring wells, push-type sampling devices (various sizes of driven well points, such as the Geoprobe© or Geopunch©), or soil gas surveys. Borings, piezometers, and

Figure 1.3 Installation of a monitoring well with a hollow stem auger (photograph courtesy of CME Company).

temporary monitoring wells (Figure 1.3) provide information about soils, geology, and ground water chemistry, flow direction(s), and rate(s).

Ground water, soil gas, and lithologic samples may be collected with push-type samplers (Figure 1.4). Push-type samplers can provide rapid results and can be cost effective in horizontally and vertically delineating contaminant plumes. Push-type samplers may also assist in placing final monitoring points. Push-type samplers may be limited to shallow applications (generally less than 70 ft) in unconsolidated sand, and their use in more lithified formations or even gravel may cause probe refusal or may bend the probe. When approved by regulatory agencies, field analysis of key constituents (versus an entire scan) may be possible.[12]

Geophysical methods may be employed either to evaluate natural hydrogeologic conditions such as depth to water table, aquifer extent, or depth to bedrock surface or to locate buried wastes, drums, or tanks.[2] Geophysical methods include airborne, surface, or downhole geophysics. Soil gas surveys may provide information about the aerial extent of contamination at a site. Indirect methods of obtaining geologic information, such as geophysical methods and soil gas surveys, may be used to augment the evidence gathered from direct field methods but should not be used as a substitute for them.[8] Data derived from monitoring points may provide information about the extent of contamination but generally not about the original mass of contaminants (Barcelona, M.J., personal communication).

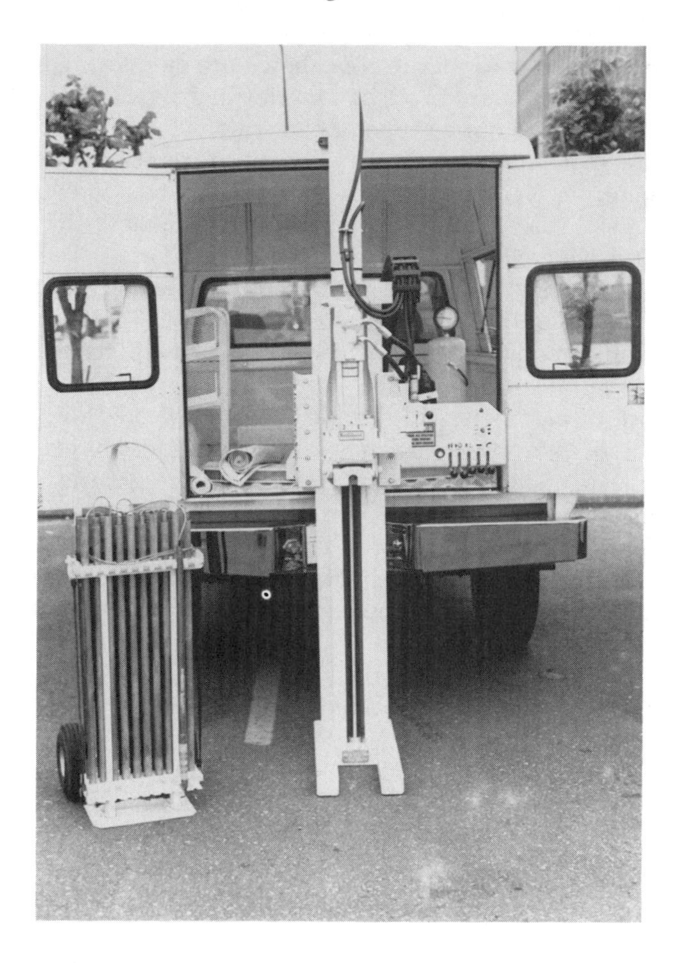

Figure 1.4 A push-type (Geoprobe®) sampling device mounted in a van.

Monitoring plans

When a decision is made to monitor a site a monitoring plan must then be developed and implemented. A monitoring plan is comprised of an environmental monitoring system (EMS), and a sampling and analysis plan which are all covered by a QA/QC component. A monitoring plan includes specifying the types, number, and vertical and horizontal placement of the final monitoring points, which may include soil water samplers, piezometers, and monitoring wells. Domestic and municipal wells may also provide information on the extent of ground water contamination.

The configuration of a monitoring system depends on an analysis of the information compiled during the previous phases of the investigation. If the site analysis indicates that complex geology underlies the site or that a large quantity of a variety of contaminants with differing physical and chemical properties are present over a large area, then more monitoring points would most likely be required than in a less complex hydrogeologic setting with a relatively small volume of one type of contaminant.

In addition to determining the number and placement of the monitoring points a monitoring plan must specify drilling methods, screen length, material and slot size, riser material, filter pack, placement of the annular seal and grout, and the well development procedure. These variables, in turn, depend on the geology and hydrogeology of the site and the physical and chemical properties of the target parameters. Chapter 3 will go into greater detail concerning monitoring point construction methods, materials, and limitations.

The choice of parameters to sample for is predicated on the expected effect on ground water by a particular use of the land. For example, target parameters for assessing agricultural impacts on the ground water would include major anions and cations, nitrogen species, and pesticides. Target parameters for landfill monitoring would include volatile organic compounds, major anions and cations, and trace metals. Sometimes, unexpected contaminants are encountered, and program objectives and the working model must be adjusted accordingly.

Sampling frequency should reflect monitoring objectives. A sampling plan and a QA/QC plan must be adequately formulated and the specified protocol adhered to, because much of the work prior to sampling was done to enable properly trained individuals to take representative samples. If sampling methodology is lacking, then much of the previous work may have been wasted. Periodically, the sampling effort should be internally and externally audited in the field. Reassessing the effectiveness of monitoring system should be a component in the overall monitoring strategy, and this can be done in annual summary reports.

The fate of contaminants in ground water and their rate of movement is determined by the physical and chemical properties of the contaminants and the subsurface conditions. Physical and chemical properties of contaminants include original mass, concentration and distribution, specific gravity, solubility, valence state, and volatility. Subsurface conditions also affect migration of pollution, and these include the types and thicknesses of potentially attenuating soils, biological activities, depth to ground water, the aquifer composition, hydraulic conductivities and horizontal and vertical extent, elevational head, presence of confining or boundary layers, and proximity to recharge or discharge areas. High-capacity pumping wells also may affect contaminant transport.

After compiling and interpreting information about a site, it may be scored based on what was found during the investigative phase. The scoring process involves integrating health and environmental risks with the degree of contamination, and allows the site to be ranked so it can be put in context with other sites which may be competing for limited resources, including time and money. A corrective action plan may be warranted which further identifies the problem and proposes and discusses various options for addressing the contamination. Corrective action options could range from

Figure 1.5 Removal of a deteriorated and leaking underground storage tank.

continued monitoring to implementing one or several possible combinations of more extensive remedial actions.

Early in the process domestic and municipal wells identified as being potentially impacted from contamination should be evaluated and sampled at periodic intervals to ensure that water supplies are protected from unacceptable levels of contaminants. If unacceptably high levels of contaminants are verified to be present in domestic wells then an interim solution must be formulated and implemented. This could include more extensive sampling, providing bottled water or installing in-house point-of-use water filtration systems until contaminant levels are reduced to acceptable levels. Drilling deeper wells into an uncontaminated aquifer or hooking up affected homes to an uncontaminated municipal water supply can serve as a more permanent solution if ground water cleanup measures are inadequate or take too long. For adversely impacted high-capacity municipal wells, installation of a water treatment system or drilling new (and possibly very expensive) deeper or more distant wells may be required.

Site remediation

Whenever possible, ground water remediation is usually initiated by first removing or reducing the source(s) of pollution. This includes removing chemical storage drums (Figure 1.5), pumping out standing pools of contaminants, relining ponds, excavating contaminated soils, reducing precipitation inflow into a site (installing a clay or synthetic cap over the site), or implementing best management practices for land application of sludge, wastewater, or agrichemicals.

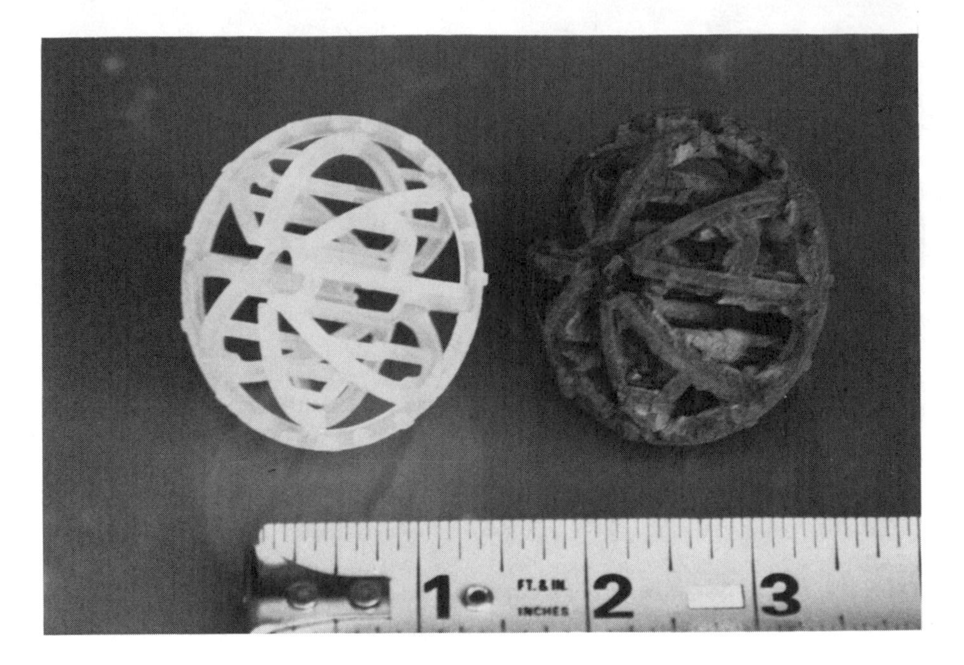

Figure 1.6 Air stripper packing media; the ball on the right was rapidly fouled by iron precipitation.

Site cleanup alternatives include excavation and burial of contaminated media in a lined, permitted landfill, installing barriers to precipitation and ground water flow, bioremediation including air sparging, bioventing, microbiologic enhancement or composting, physically or chemically treating the soils and/or the ground water, allowing natural dilution, dispersion, and attenuation to occur, or combinations of the above. The treatment method selected may first need a bench scale or pilot project (treatability study) to ensure that the proposed method will work in the specific hydrogeologic context.

Barriers to ground water flow include barrier wells, which alter the natural ground water flow regime; utilizing seepage trenches or drain tiles, which intercept a portion of near-surface contaminant flow; or installing solid or slurry walls, which retard ground water flow.

For some sites, such as unlined dumps or landfills, removal of the original source of contamination may not be possible. Ground water contamination may be reduced at unlined dumps or landfills by installing relatively impermeable clay and/or synthetic caps to reduce infiltration of precipitation, installing venting systems, or installing seep collection and treatment systems. Nonremoval contamination situations are handled on a case-by-case basis, using the site's corrective action plan if one was warranted.

Utilizing aerobic and anaerobic bacteria may also be a viable treatment option. The optimal strain of bacteria for bioremediation depends on the concentration and nature of the contaminant(s). Bacteria can be utilized at land application sites (sludge- or petroleum-contaminated soils) and in above-ground biologic reactors (ponds, tanks, lagoons). They can be introduced into the aquifer, or existing suitable bacteria populations within the aquifer matrix can be enlarged by the addition of nutrients or oxygen (*in situ* treatment).

Physical and chemical treatment systems for volatile organic compounds include bioremediation, thermal desorption, air stripping, carbon adsorption, incineration, and chemical oxidation. Treatment for inorganic contaminants include stabilization/immobilization, chemical addition, removal of suspended solids, ion exchange, reverse osmosis, electrodialysis, distillation, and biologic treatment.[7] Landfilling of hazardous wastes should be reserved for contaminants which do not lend themselves to alternative methods of treatment due to their physical or chemical properties.

Regulation of contaminated sites

The components that drive ground water investigation and remediation include hydraulics, contaminant concentration, distribution, mobility and toxicity, relative impacts, and technological, regulatory, political, sociological, and monetary considerations. The costs associated with remediation can be extensive. A thorough site characterization that incorporates an adequate ground water monitoring component is essential to effectively assess and attempt remediation at a contaminated site and to reduce the associated expenses.

In terms of ground water contamination, the three components necessary for a threat to human health or the environment are a source of contamination, a pathway, and a receptor(s). In some instances ground water remediation may not be practical or feasible due to technical or monetary limitations, low relative health and/or environmental risks, or other considerations. Attempts at remediation are sometimes instituted, even at a relatively low benefit-to-cost ratio, due to public concerns. Programs to assist in educating the public about environmental issues help to ensure that solutions formulated and implemented are protective and feasible and reflect technical and monetary constraints.

Several federal and state laws in the United States have been established to protect ground water resources. Some states have implemented more rigorous requirements than those required by the federal government. Table 1 summarizes federal laws designed to protect the ground water.

States can take primacy over ground water protection and cleanup programs and can require more stringent standards than those mandated by federal laws. It is usually in the best interest of a state to meet or exceed minimum federal requirements for ground water protection to protect human health and the environment. Also, the federal government can impose fines (withhold funds) or take over part or all of a state's environmental program if minimum requirements are not being met.

Table 1 Federal Laws Concerning Ground Water Contamination and What the Laws Address

Federal Law	Date Enacted	Summary of the Law
Clean Water Act	1972	Regulates discharges into all navigable waters in the United States Provided funding for construction of municipal sewage treatment plants
Safe Drinking Water Act	1974	Established standards for ensuring the safety of drinking water Regulates injection wells Protects sole source aquifers
Toxic Substance Control Act	1976	Identifies and controls chemical products that pose an unreasonable risk through their manufacture, distribution, processing, use, or disposal (e.g., PCBs)
Resource Conservation and Recovery Act (RCRA)	1976	Established guidelines for managing hazardous wastes (Subtitle C) Established guidelines for managing solid wastes (Subtitle D) Regulates certain underground storage systems
Surface Mining Control Act	1977	Requires ground water monitoring before, during, and after mining activities
Comprehensive Environmental Response, Compensation and Liability Act (CERCLA) (otherwise known as Superfund)	1980	Facilitates cleanup of wastes resulting from accidents in transporting hazardous wastes Makes responsible parties pay for cleanups Facilitates cleanup at sites where ownership cannot be determined
Superfund Amendments and Reauthorization Act (SARA)	1986	Increased trust fund from $1.6 to $8.5 billion Set goals and deadlines for remedial investigations and preliminary assessments Strengthened enforcement Increased state involvement Mandated emergency planning

Prevention of ground water contamination

In addition to attempting remediation of existing ground water contamination problems, various federal, state, and local efforts are underway to prevent contamination problems before they occur. Using less hazardous solvents, recycling of household wastes, more stringent laws that regulate underground storage tanks, imposition of civil or criminal penalties for noncompliance with environmental regulations, utilizing a permit system to address land application of various waste groups, and cradle-to-grave manifest systems for dealing with hazardous wastes are examples of strategies to reduce waste volume and improper disposal of wastes and the subsequent potential for ground water contamination. Proper site selection, engineering, and monitoring of new waste disposal sites also serve to reduce ground water contamination before it occurs.

The permit system

Typically, a facility, industry, or waste generator in the United States operates under a permit system. The permit stipulates the minimum requirements that the permittee must meet in order to be in compliance. Not meeting the requirements of the permit can trigger enforcement actions, which can range from verbal or written warnings, fines, and closing down the operation to possible civil or criminal penalties. The permit can stipulate monitoring requirements (including air, surface, and ground water monitoring), waste handling procedures, where waste products must go, maximum discharge rates of parameters of concern, and reporting requirements including monthly, quarterly, and yearly submittal of the required data.

Data should be presented in a manner that lends itself to trend analysis. Optimally, data should be transferred electronically so that statistically based trend analysis can be facilitated. Permits help to track the types and quantities of wastes that can potentially impact the atmosphere, surface water, and ground water, and are thus tailored to the particular wastes to be regulated.

When a site is permitted (for example a landfill, cannery, or a municipal wastewater treatment facility), intervention limits may be stipulated in the permit that are percentages of the drinking water standard for each parameter. Intervention limits may be specified in a permit to provide some lead time in order to reduce the parameters of concern before they become a pronounced health or environmental risk off the facility's property.

Upgradient wells are required to determine the extent to which a site is impacting the ground water. If intervention limits are exceeded in downgradient wells and the upgradient wells are unimpacted, then enforcement actions, which may include financial penalties, may be imposed in addition to "fixing what's broke". The extent or flagrancy of the permit violation should dictate the magnitude of enforcement action(s). Compliance monitoring under a permit also serves to provide a degree of protection and a feedback loop for a facility's owner or operator. If a permittee can demonstrate compliance, then some measure of relief from public concerns and from legal actions can be afforded. If a permittee is out of compliance, then specific measures can be taken to improve the system's performance and return it into compliance.

The permit system has sometimes been disparagingly referred to as "a permit to pollute" and to a certain extent this may be true. The permit system, however, does provide a mechanism for evaluating a site's performance and gives some regulatory control over its operation. Part of the process of permitting a facility should require that individuals who manage wastes are adequately trained to do the job effectively. An ongoing operator training program is needed for optimal waste management.

Unpermitted land use activities that often impact the ground water include leaking underground storage tanks, spills, "midnight dumpers", nonconforming on-site septic systems, and many agricultural activities. Leaking underground storage tanks should be removed, the site cleaned up, and replacement tanks should be designed to minimize the risk of future leaks.

For spills and "midnight dumpers", cleanup should be mandated, and civil and possibly criminal penalties could be assigned if the perpetrators are caught and there was criminal intent. For on-site septic systems, better siting, design, and construction can assist in minimizing their collective impacts. Best management practices for agriculture are being encouraged for the storage and application of pesticides and fertilizers, including manure. More regulation of the agricultural industry may be forthcoming.

Ground water sampling

Ground water sampling is a vital component of ground water monitoring. When taking ground water samples, consider that the documentation and protocol used may be scrutinized hostily in a court of law. A ground water sampling protocol should include taking proper field parameters, using proper purge and sample acquisition techniques, ensuring that a robust QA/QC plan is adhered to, and providing adequate documentation. All of the other (expensive) components of investigation and remediation may be rendered worthless if proper samples are not taken. Perhaps one of the weakest links in ground water monitoring is sampling. Table 2 illustrates how a relatively small error can produce very large results.

Table 2 Results of Errors in Sampling

One Part Per Million	One Part Per Billion
1 minute/2 years	1 second/32 years
1 teaspoon in 1,300 gallons	1 teaspoon in 5 railroad tanker cars
1 sq. ft./2.3 acres	1 sq. in./ 4 sq. miles
1 bad apple/2,000 barrels	1 bad apple/2,000,000 barrels
1 inch/16 miles	1 inch/16,000 miles
1¢/$10,000.	1¢/$10,000,000.

Figure 1.7 Sarah Elizabeth, age 6.

The primary goal for ground water sampling is to obtain representative samples and field parameters. Adequate design and supervision are needed to minimize errors introduced through improper methods of monitoring well construction, development, purging, and sample acquisition. The following chapters are intended to reduce these types of errors and to maximize the performance of the monitoring system.

References

1. ASTM, *Standard Practice for Environmental Site Assessments: Phase I Environmental Site Assessment Process,* E 1527-93, American Society for Testing and Materials, Philadelphia, PA, 1993.
2. Benson, R.C., Remote sensing and geophysical methods for evaluation of subsurface conditions, *Practical Methods of Ground-Water Monitoring,* Lewis Publishers, Chelsea, MI, 1991.
3. Keith, L.H., *Environmental Sampling and Analysis: A Practical Guide,* Lewis Publishers, Chelsea, MI, 1991.
4. Kerr, J., Investigation and remediation of VOCs in soil and ground water, *Environ. Sci. Technol.,* 24(2), 1990.
5. Madsen, E.L., Determining in situ biodegradation, *Environ. Sci. Technol.,* 25(10), 1991.
6. Nelson, J.D. and Ward, R.C., Statistical considerations and sampling techniques for ground-water quality monitoring, *Ground Water,* 19(6), 1981.
7. Nyer, E. K., *Ground Water Treatment Technology,* Van Nostrand, Reinhold, New York, 1985.

8. USEPA, RCRA Ground-Water Monitoring Technical Enforcement Guidance Document, U.S. Environmental Protection Agency, Washington, D.C., 1986.
9. USEPA, Data Quality Objectives for Remedial Response Activities, EPA/540/G-87/003, U.S. Environmental Protection Agency, Washington, D.C., 1987.
10. USEPA., Test Methods for Evaluating Solid Waste, SW-846, U.S. Environmental Protection Agency, Washington, D.C., 1992, chap. 1.
11. Ward, R.C., Water Quality Monitoring — A Systems Approach to Design, Int. Symp. on the Design of Water Quality Information Systems, June 7–9, 1989.
12. Spittler, T.M., Field Instrumentation and Needs of U.S. EPA, MIT/Marine Industry Collegium Opportunity Briefs, No. 61, Report No. MITSG 92-17, October 26–27, 1992.

chapter two

Soil water monitoring devices

"Water Water Everywhere Nor Any Drop to Drink."
The Rime of the Ancient Mariner,
Samuel Coleridge, 1772–1834

The unsaturated zone

Water entering the ground from the earth's surface moves downward, due to the influence of gravity and unsatisfied forces in the soil matrix, through the vadose zone toward the saturated zone. The vadose zone (from the Latin word *vadosus* meaning shallow) begins at the earth's surface and extends downward to the first principal water-bearing aquifer. The vadose zone has been subdivided into three zones: the soil zone, the intermediate vadose zone, and the capillary fringe.[8] Flow in the vadose zone changes from varying degrees of partial saturation to episodes of preferential saturated flow.[6]

It is sometimes desirable to obtain a soil water sample in the unsaturated zone to fulfill monitoring objectives such as providing an "early warning" network or because of difficulties in siting monitoring wells in some geologic media such as fractured bedrock or a clayey regolith. Monitoring the vadose zone, in addition to using monitoring wells, may be necessary to resolve questions about contaminant migration. Pollutants in the vadose zone may or may not travel at the same rate as water, but the travel time of water represents an upward limit on the mobility of some pollutants, notably trace metals.

Soil water samplers

The term *lysimeter* has been used by soil scientists to refer to a basin filled with soil which is used to measure quantities of water taken up by plants, evaporated from the soil, and lost by deep percolation.[1] The term has also been used in reference to devices that collect soil moisture, which should more properly be termed soil water samplers. The two major types of soil water samplers are vacuum and percolate types. There are advantages and disadvantages with both types, and which sampler to use depends on the objectives of the monitoring plan.

What is commonly referred to as a vacuum lysimeter (also called a suction or tension lysimeter) utilizes a vacuum to draw water into the sample collection vessel and thus collects dynamic or static interstitial water and

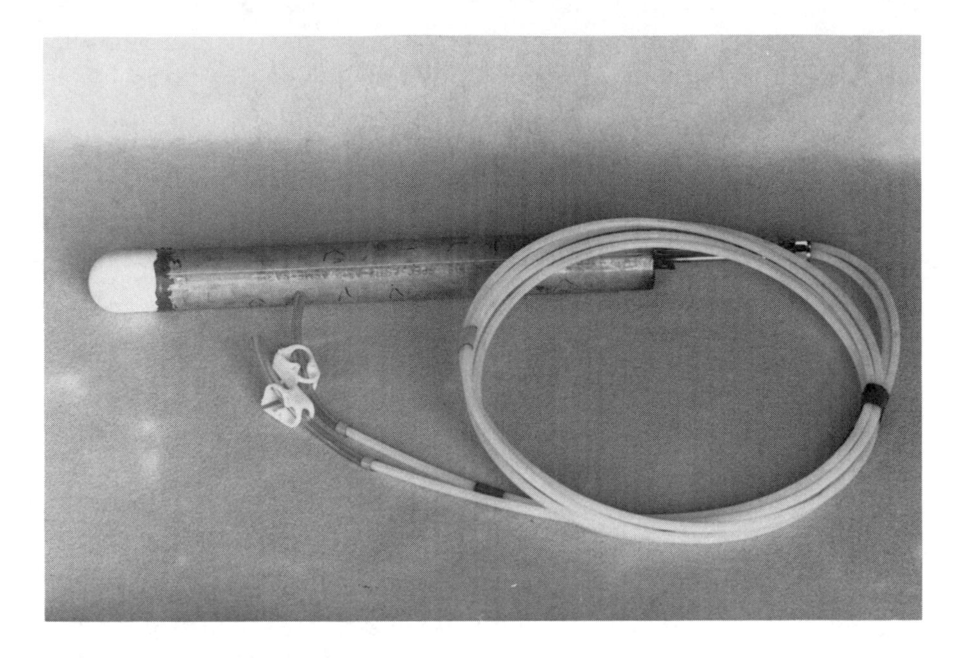

Figure 2.1 A ceramic cup pressure/vacuum soil water sampler and sampling lines.

water adhering to soil grains or clasts. Percolate soil water samplers, including the pan and wick types, rely on gravity and/or capillary action to intercept both matrix and preferential (macropore) flow from precipitation or irrigation wetting fronts. Macropores include worm channels, cracks, or root channels. Due to the differences in residence times related to how a sampler type derives its sample, a comparison of soil water chemistries between the vacuum and percolate types is difficult.[12]

Pressure/vacuum soil water samplers

A pressure/vacuum soil water sampler consists of a half-hemispheric porous ceramic, PTFE (Teflon™), or stainless steel tip attached by epoxy glue or flush threading to a PVC or stainless steel tube (Figure 2.1). The tube is typically 1.5 to 2 in. in diameter and varies from 4 in. to over 18 in. in length. An inlet and outlet tube exit from the back plate of the device, and two chemically inert (usually PTFE) tubing lines run from the device to the ground surface. To obtain a sample a vacuum is applied to the sampler, and sufficient time is allowed to elapse so soil water is drawn into the sampler. A pressure or a vacuum is then applied to one of the sampling lines, and as a result the sample is brought to the surface (Figure 2.2).

Pressure/vacuum samplers are limited to a maximum depth of 15 m. When extracting a sample with a pressure in excess of one atmosphere, damage may occur to the tip, or the sample may be forced back through the tip and into the soil on application of pressure.[24] High-pressure vacuum

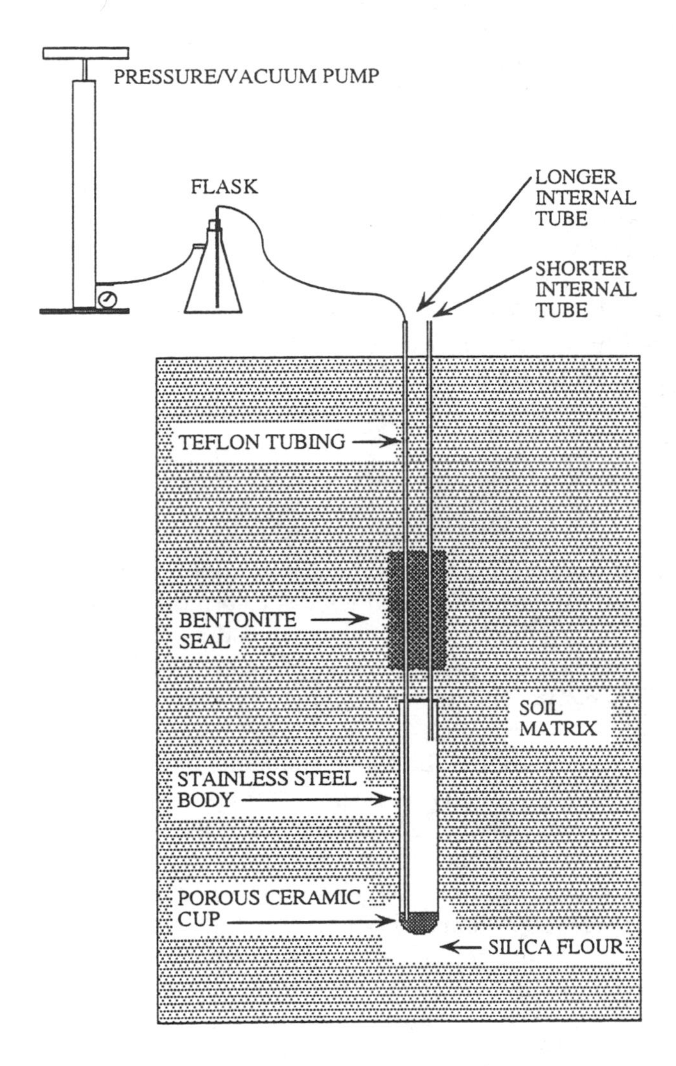

Figure 2.2 A pressure/vacuum soil water sampler and the equipment used to obtain a sample.

samplers utilize check valves and a separate sample reservoir in the sampler design to prevent this. High-pressure samplers have been used to depths of over 90 m.

The literature has cited limitations associated with suction samplers including:

- The tip of the sampler may become plugged by microorganisms or fine material[16]
- Samples may be altered due to redox or pH changes[14]
- Samplers may add or subtract chemical constituents by leaching from or adsorption to the sampler[3]

- Reservoirs ("dead space") of between 34 to 80 mL of water may remain in the sampler and cannot be extracted. Furthermore leaks in the sampler or the lines may develop, and volatile organic compounds (VOCs) may be lost[9]
- A sufficient sample volume may be difficult to obtain[14]
- Natural flow paths in the vadose zone may be altered during installation and/or sampling[17]
- Obtaining a good hydraulic connection between the sampler and surrounding soils may be difficult, especially in coarser materials, and freeze/thaw cycles may cause the loss of hydraulic contact and may fracture ceramic cups[9]
- The time necessary to extract a sample may exceed sample holding times[21]
- Upwards of 100 samples may be required to obtain a 70% confidence level[20]

If a nonconstant (one evacuation per sample) vacuum is utilized for obtaining a soil water sample, the device is referred to as a falling head soil water sampler, since the vacuum decreases as the sample is drawn into the sampling body. The decrease in vacuum over time from a falling head sampler may skew the chemistry in soil water from wetting fronts derived from precipitation or irrigation events. This is because the soil water at the front of a wetting front may be more concentrated with major ions, trace metals, and volatile and semivolatile organic compounds than the back end of the wetting front, and most of the vacuum in the soil water sampler may be expended by the front end of the wetting front.

The wetting front loading phenomenon can be diminished in vacuum soil water samplers by utilizing either an above-ground vacuum tank with a vacuum regulator or a continuously running pump to keep the vacuum in the sampler constant. In addition to preventing the wetting front from skewing the sample, a constant vacuum provides more sample volume than a nonconstant vacuum[16] and VOCs are lost if the vacuum is intermittently reestablished.[10] By keeping a continuous vacuum at 10 centibar (cbar) or less, the sample will more likely represent macropore flow.[2]

Several types of vacuum soil water samplers are available including those with tips made of ceramic material (including alundum), stainless steel, and PTFE. Vacuum samplers made of PTFE were reported to have an operational range of 0–7 cbar of suction, which may be too low for most sampling objectives. Furthermore, PTFE samplers did not hold a vacuum for an appreciable length of time.[14] Cleaned PTFE samplers may leach higher levels of copper, iron, and chromium than cleaned ceramic or alundum cups.[7]

Ceramic cup samplers are composed of 55% Al_2O_3, 35% SiO_2, and minor amounts of Fe_2O_3, TiO_2 CaO, MgO, Na_2O, K_2O, and SO_3. Alundum cups are composed of 90% Al_2O_3, with minor amounts of SiO_2, Fe_2O_3, and TiO_2.[7] Samplers with ceramic tips reportedly can be used for over three weeks without an appreciable drop in vacuum, but the tip may adsorb trace metals (cadmium, cobalt, chromium, zinc) and leach several milligrams of calcium, magnesium, sodium, bicarbonate, and silica, even after cleaning with dilute hydrochloric acid.[3] Ceramic cups from different manufacturers have been reported to vary significantly in potassium, calcium, magnesium, and sodium concentrations,[18] and this could result in differential leaching between samplers from different manufacturers.

Samplers with tips made from stainless steel obtained more sample volume at quicker rates than those with ceramic tips in near-saturated conditions in finer soils.[14] Stainless steel membrane samplers adsorb fewer trace metals than ceramic samplers,[13] and if properly constructed and cleaned prior to installation they should not leach appreciable amounts of major ions.

Vacuum samplers with ceramic tips should not be used for major ion sampling if the total dissolved solids concentration in the soil water is less than 500 mg/L because major ions leaching from the tip have a proportionally larger effect at lower soil water ion concentrations. Ceramic tip samplers should not be used for acquisition of trace metals due to adsorption of trace metals by the tip.[3]

Silica flour is sometimes used to increase the hydraulic contact between the soils and the sampler tip and also to reduce plugging problems. Silica flour has been shown to adsorb trace metals on a pH dependent basis; it should not be used for samplers designed for trace metal acquisition.[13]

Percolate soil water samplers

In a well-structured soil most of the water movement is through macropore flow,[21] and water and solute transport have been found to be dominated by the presence of macropores.[23] Percolate types of soil water samplers (referred to as pan, free drainage, or zero tension samplers) yield samples in a structured soil that may better reflect the chemistry of water migrating toward the water table since they intercept macropore flow. Other advantages of percolate soil water samplers include larger sample volumes, no need for a continuous vacuum, less potential for losing volatile compounds, and greater confidence with fewer samples, as a 70% confidence level is obtained with half the number of samples required for tension samplers.[20] Percolate samplers are also less likely to alter natural flow conditions, especially if large pans are used.

Percolate samplers, however, are usually more difficult (and costly) to install properly than vacuum samplers, and the pan types will only function when the soil moisture is greater than field capacity. Figure 2.3 is a flow chart that can assist in determining the suitability of soil water samplers at a particular site. Surrogate sampling (sampling a less reactive parameter in lieu of a more reactive one) may be considered in a soil water sampling network.

Limitations inherent to soil water samplers

Flow direction(s) associated with a soil water sample may be uncertain. It is possible that soil water may be in chemical disequilibrium at the sampling depth in relation to water that migrates to the first useable aquifer. Regulatory

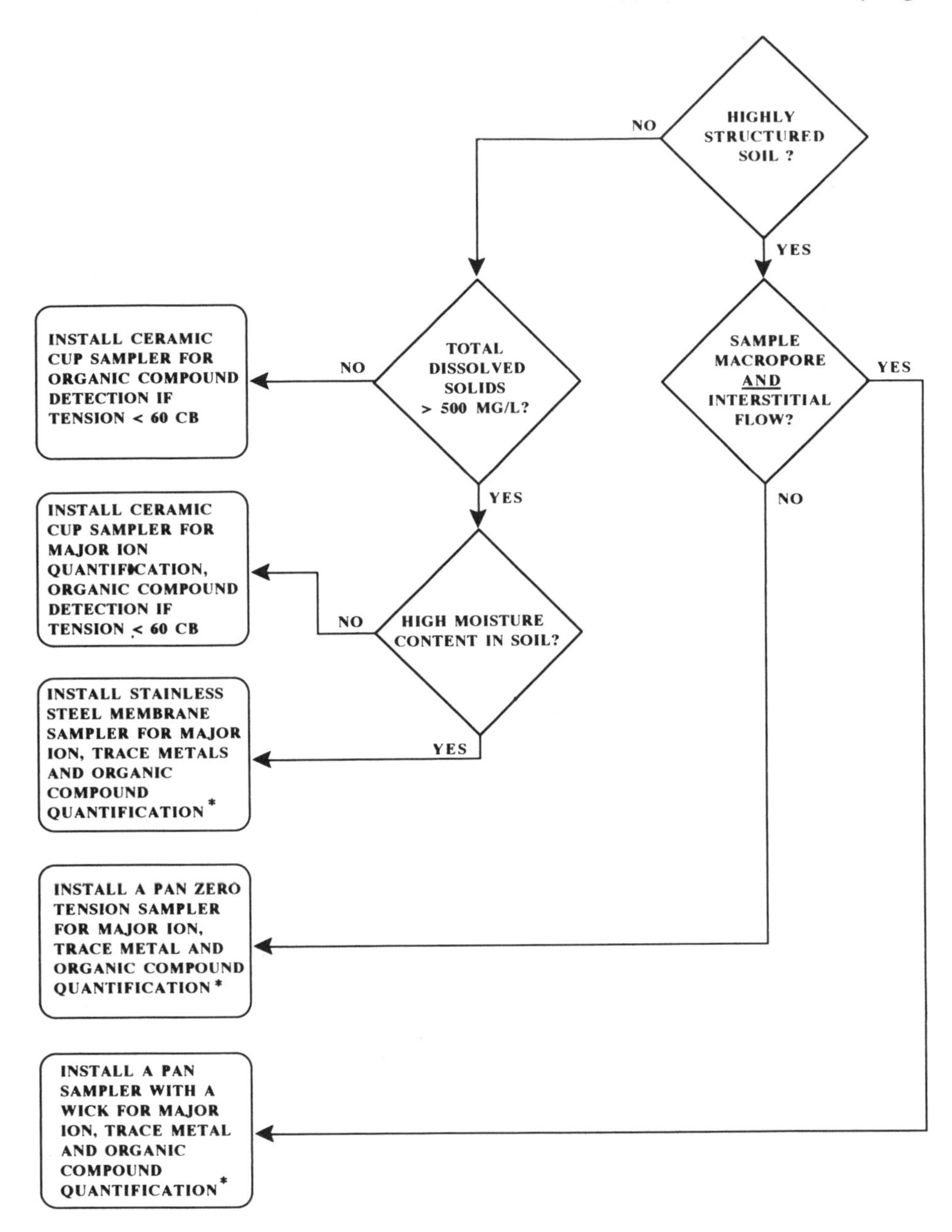

Figure 2.3 Decision tree for determining soil water sampler suitability.

agencies may view soil water (which is not used for drinking water purposes) differently than ground water. Some sites may be unsuitable for soil water sampling because the water table is too close to the ground surface to install a soil water sampler, the soil is not amenable to acquiring a sample within required holding times (too dry or too tight), or the number of samplers required to meet statistically based levels of confidence precludes their installation.

Site operations may be hindered by the installation and operation of soil water samplers and cause sample bias. An above-ground locked riser pipe, where the sampling lines are stored between sampling events, or the use of a site shelter for percolate samplers may interfere with site operations. This may bias the monitoring regime, as the same land use activities may not be occurring around the lines or shelter as on other areas of the site. In addition, excessive traffic around the sampler may reduce the amount of infiltration in proximity to the sampler. These alterations could lead to a difference in chemistry between site locations.

Shelters provide relatively easy access to the sampler, and if the sampler fails it can readily be inspected and corrected. The use of long trenches to run sampling lines to less obtrusive sampling stations or making the obstruction less pronounced can assist in reducing these potential problems. Burying the lines with rust-resistant lengths of chain or a steel plate (smaller pieces encased in plastic bags), using a measuring tape for determining relative distances, and then using a metal detector to find the lines at the onset of sampling also could be considered.

Placement of soil water samplers

Depending on the objectives of the monitoring plan, most if not all of the following should be known before installing soil water samplers in order to determine the applicability of sampling at a particular site or to assist in determining the types of devices and number of samples required to meet statistically based confidence levels:

- Soil texture
- Soil structure
- Soil pH
- Percent organic matter
- Thickness of soil horizons
- Type of lithology
- Depth to bedrock
- Depth to the seasonal high-water table and presence of perched water
- Cation exchange capacity (CEC) in the top soil
- Horizontal and vertical hydraulic conductivity
- Anticipated atmospheric and artificial hydraulic loading rates
- Maximum infiltration rates
- Extent of anisotropy in horizons above and within the sampling zone
- Target parameters, holding times, and estimated time required to obtain a sample
- Sample volume required from the sampler
- Specific conductance/soluble salts measured on a saturation extract of the soils

These variables are obtained by compiling existing information, field methods (including borings, tensiometers, infiltrometers, piezometers) and laboratory determinations. The extent to which the vadose zone is adequately characterized before a soil water sampler network is installed determines the chances that the network will succeed. After all necessary information has been obtained and compiled, a determination needs to be made on the types and numbers of soil water samplers to install at a site. Even with the above information, the inherent uncertainties in the vadose zone may make the number and horizontal and vertical placement of soil water samplers difficult to ascertain. Mathematical models can assist in estimating sampler spacing and the time required to obtain a sample volume.[5]

The use of tensiometers or electrical resistance (gypsum) blocks in conjunction with soil water samplers is usually warranted as they can measure the degree of soil saturation prior to sampling, can assist in determining how the water in the sampler was derived (for example, from the last precipitation event), or can be used to determine when a vacuum should be applied to a vacuum soil water sampler to obtain a sample within specified holding times.

Preparation of pressure/vacuum samplers

An appropriate and approved safety and health plan must be prepared and used for all laboratory and field activities (see Chapter 6). Extreme caution must be exercised when using strong acids. The health and safety plan for preparing soil water samplers must include the use of safety equipment such as acid-resistant safety goggles, acid-resistant gloves, and appropriate clothing, as well as provisions for adequate ventilation, a suitable first aid kit, first aid training, immediate access to running water, and an emergency wash station. All cautions and warnings associated with products and outlined in the health and safety plan governing sampler leak checking, cleaning, installation, and sampling must be read and followed.

Antecedent conditions have a critical effect on the performance of soil water samplers. If a vacuum sampler is selected, it must be leak-checked and thoroughly cleaned prior to installation. The air pressure required to force air through the porous tip of a soil water sampler that has been thoroughly wetted is called the *air entry value* or *bubbling pressure*. The bubbling pressure for low-flow ceramic cups is typically between 250 to 300 cbar, and for stainless steel samplers between 25 to 30 cbar. The bubbling pressure for each sampler type, which is provided by the manufacturer, is equal to the maximum suction that can be applied to the sampler before air entry occurs.

Following the manufacturer's instructions, the first step in preparing a soil water sampler for installation is to check it for leaks. Leak-checking a ceramic cup soil water sampler is typically initiated by pulling approximately 500 mL of deionized water through the tip at 80 cbar of suction. The tip is then submerged for a minimum of 2 hours in deionized water. All the water should then be removed from the sampler. Next, the sampler is placed in an aquarium filled with deionized water and pressurized to just below the bubbling pressure. All fittings, connections, and the tip should be carefully checked for leaks. Even small leaks can render the unit dysfunctional.

After inspecting the sampler for leaks, it should be cleaned. The protocol used for cleaning a soil water sampler depends on the type of sampler and the target parameters. The exact cleaning procedure proposed for soil water samplers for a given site should be specified in the approved quality assurance/quality control (QA/QC) plan and rigorously followed.

Typically, cleaning a (dry) vacuum soil water sampler prior to installation is undertaken as follows:

1. For ceramic cup samplers it has been recommended that approximately 500 mL (70 pore volumes) and for alundum cup samplers, 750 mL (60 pore volumes) of reagent grade 1 *N* hydrochloric acid be gravity-drained through the cup to reduce major ion concentrations.[7] When using a stainless steel membrane vacuum sampler, 1 to 10% of reagent grade nitric acid should be gravity drained through the tip of the sampler for up to an hour.[19] Acid may corrode valves, so the above flushing procedure should be performed prior to tip attachment, if possible.[2]
2. After properly removing and disposing of all the acid in the sampler, rinse the outside of the cup and body thoroughly with deionized water. After rinsing the outside of the sampler, gravity drain approximately 15 to 20 L of deionized water through the sampler cup. When the specific conductance of the deionized water going into the cup has a less than 2% difference from that coming out of the sampler and the pH of the output water equals that of the input water, the cups are adequately rinsed.[7]
3. Soil water samplers used to collect volatile or semivolatile compounds may need to have the above cleaning procedure applied with an additional acetone rinse(s) prior to the final deionized water rinse.

As per the QA/QC plan, deionized water aliquots should be taken through the tips of a predetermined number of samplers (no less than 1 sampler in 15 or 1 per cleaning batch) prior to placement and analyzed for the same target parameters to ensure that the devices are acceptably clean. These samples are called *equipment blanks.* Enough lead time for analysis turnaround of the equipment blank(s) prior to installation must be provided. The use of a new pair of acid-resistant disposable plastic gloves for cleaning each sampler and placing samplers in clean plastic bags after cleaning are recommended to reduce the potential for contamination.

Installation of pressure/vacuum samplers

In accordance with the approved health and site safety plan and OSHA Regulation 1926.651, all utilities (including gas, electric, phone, water and sewer) and all underground pipeline companies must be given adequate notification prior to drilling or trenching. Installation of vacuum samplers is usually accomplished by either digging a trench (properly shored against cave-in) and installing the sampler at a 45° angle from the horizontal to reduce soil perturbations,[22] or drilling a hole into the soil to the specified depth, which is usually just below the rooting zone (top 5 ft).

The use of disposable plastic gloves for soil water sampler installation is recommended in order to reduce the potential for contamination of the sampling device. Ceramic cup samplers must be installed with thoroughly wetted tips as they will not hold a vacuum if dry.[10] Adequate wetting is accomplished by placing the tip of the sampler in deionized water for approximately 30 min prior to installation. After the hole or trench is dug, a wetted slurry of silica flour or native material is placed around the tip of the sampler to enhance hydraulic conductivity between the tip of the sampler and the soil. Silica flour should not be used when sampling for trace metals. Since vacuum samplers may float in silica flour, either filling the sampler with deionized water or using rigid risers is recommended when using silica flour.[2] Freezing a silica flour slurry to the tip has also been used to install vacuum samplers.

A tensiometer or a gypsum block should be installed along with the sampler. This is done to establish that enough moisture is available to obtain a sample within specified holding times, to coordinate sampling with specific wetting fronts, and to assess the performance of the sampling device. Small-diameter tensiometers equipped with pressure transducers and time domain reflectometry probes have been found to be more sensitive than large-diameter bourdon gauge tensiometers.[15]

The longer and shorter (internal) sampler tubing lines should be labeled as such at the surface end of the tubing for later use in applying pressures and vacuums. The same sequence of soil as was originally removed is carefully placed over the top of the device and is tamped as the soils are replaced to reduce soil perturbations. The sampler should be accurately located horizontally and vertically on a site map and, depending on monitoring objectives, should be surveyed in with horizontal control. Optimally, a year should elapse prior to using the sampler to allow it to come into equilibrium with the soil and soil water constituents.

Sampling pressure/vacuum samplers

To obtain a falling head type of sample, a vacuum is introduced from the ground surface via one of the PTFE tubing lines, with the other tubing line being clamped off. After introducing the vacuum, the vacuum line is then clamped off. The amount of vacuum to apply is provided by the manufacturer; otherwise, a vacuum of 25 to 30 cbar for stainless steel membrane samplers and 40 to 60 cbar for ceramic cup samplers are typical. The maximum vacuum that can be applied to a vacuum soil water sampler is 100 cbar (1 atmosphere). Prior to introducing the vacuum, a test pressure or a vacuum should be applied to one line and air flow should be observed in the other line to ensure that the lines are not obstructed.

The amount of time necessary to draw enough water into the sampler is a function of the amount of soil moisture, the type of soil and sampler, and the amount of vacuum initially applied. It is not uncommon to introduce a vacuum into a sampler and wait several days before going back to obtain the sample. The first two samples after installation should be discarded to allow initial adsorption of cations in the soil water to come into equilibrium with the sampler tip.

Pulling a vacuum just prior to or shortly after a precipitation or irrigation event and using tensiometers or electrical resistance blocks to provide soil moisture information can reduce the time required to obtain a sufficient sample volume to within sample holding times. Vacuum soil water samplers cannot extract a sample if the matrix potential is greater than 1 atmosphere, and a sample is difficult to obtain if the matrix potential is above 60 cbar.[9] This condition is usually found in dry, coarser soils.

It is recommended that before extracting a sample for analysis the sampler be purged of native soil water.[13] This is accomplished by pulling a vacuum into the device, obtaining soil water, and discarding it immediately before pulling a vacuum to obtain a sample. Vacuum samplers may be purged with nitrogen gas prior to sample acquisition to reduce the potential for sample alteration due to interaction with atmospheric gases.

An adequate amount of time should be allowed to elapse based on the soils, the sampler, the degree of saturation, and previous experience. Soil water is then brought to the surface via one of the sampling lines. This is accomplished by either forcing compressed nitrogen gas down one line connected to the shorter length of rigid tubing inside the body of the sampler, or by exerting a vacuum on the sample line connected to the longer tube inside the sampler body .

If volatile organic compounds (VOCs) are to be collected, then (pure) nitrogen gas must be used to push water out of the sampler body instead of using a vacuum. This is done to prevent VOCs from being drawn out of solution by the vacuum. The nitrogen gas pressure must be only slightly higher than the pressure required to lift the sample or else bubbling and loss of dissolved gases may occur. VOC samples must be delivered from the tubing directly to laboratory precleaned 40-mL borosilicate purge-and-trap vials in a manner similar to obtaining a ground water sample (see Chapter 6). For extraction of major ions from a sampler body, the use of a vacuum and an adequately cleaned side-arm Erlenmeyer flask may be acceptable. All sampling procedures must be done as per the formal, written QA/QC plan.

Troubleshooting pressure/vacuum samplers

Several possible reasons could account for the failure of a vacuum sampler to obtain a soil water sample. Reasons for failure include lack of enough soil moisture to obtain a sample, poor hydraulic contact between the tip and soil, leakage of the sampler or lines, a fractured tip, or clogging of the tip or lines. If a tensiometer or gypsum block was installed with the sampler, then the soil moisture content should first be determined to ensure that it is possible to obtain a sample. If enough soil moisture is available for sample acquisition, then apply a gentle pressure or vacuum to one of the sample lines. If air movement is detected in the other line then the sampling lines are most likely intact.

If enough moisture is available and the lines appear to be intact then, for ceramic cup samplers, apply 80 cbar of suction and observe the decay of suction over time. A decay in vacuum over a period of hours is considered normal.[2] An almost instantaneous decrease in suction is due to leaks in the sampler, and a decline over a period of minutes may be due to air entering the sampler in dry soils (refer to the tensiometer or gypsum block). If the suction does not decay, or does so over several days, the tip may be plugged, or in the case of a high-pressure sampler, a check valve may have failed.

Drawing a known volume of deionized water from the surface through the longer internal line and out the shorter internal line of the sampler into a container at ground surface will ascertain tip plugging or fracturing in a pressure/vacuum sampler. If the input and output volumes are close, then the tip may be plugged. A plugged tip may be reconditioned by gently forcing deionized water from the surface, under pressure, out through the tip. This may introduce unrepresentative water into the zone of influence of a pressure/vacuum sampler, and this procedure cannot be used for checking potential tip plugging or fracturing of high-pressure samplers.

To reduce the potential for failure due to freezing, the sampler and lines should be purged of soil water before the onset of freezing weather. Prior to pulling a vacuum to obtain a sample, the lines should be purged of soil water by introducing a vacuum in one line with the other line left open. A comprehensive, step-by-step installation, sampling, and analysis plan coupled with a QA/QC plan should be formulated for each site, and the sampling personnel should be familiar with the detailed sampling and troubleshooting procedures.

Percolate soil water samplers

Soil structure is a field term which describes the arrangement and aggregation of soil solids. The structure of a soil is derived from parent material; climate; soluble salts; organic matter; downward migration of clay, iron oxides, and lime; and biologic action. Soil structure varies over time and space and includes single-grained, massive, and aggregated structures.[11] The observable forms of soil aggregation include plate-like, prism-like, block-like, and spheroidal structural types.

Macropore flow derived from worm channels, cracks, and root channels is dependent on soil structure. The optimal type of soil water sampler to intercept macropore flow is a percolate sampler. There are two main types of percolate samplers — pan and wick. As percolate samplers are not readily available from vendors they must be fabricated by those who wish to install them. Pan percolate samplers must be constructed of nonporous materials that will not alter the chemistry of the sample. Sheet metal and glass blocks,[21] aluminum,[12] or plastic[20] have been used to fabricate pan samplers. Figure 2.4 is a diagram of a typical pan percolate sampler.

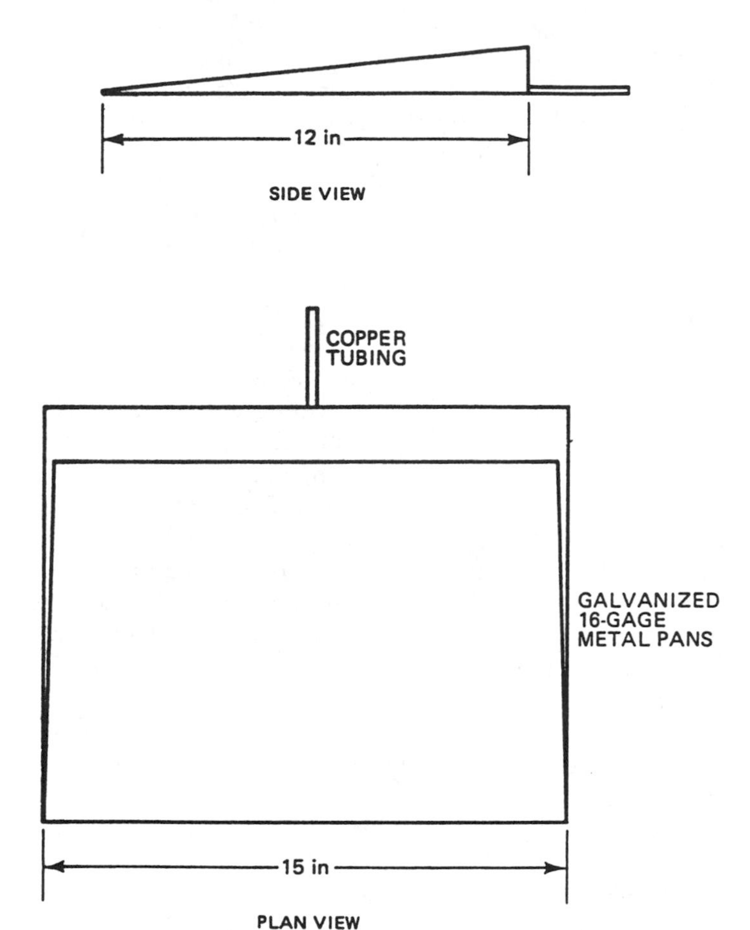

Figure 2.4 Pan percolate soil water sampler.

Essentially, a pan percolate sampler is a nonreactive pan, filled with an inert packing material (sand or polypropylene pellets) or material excavated from the emplacement tunnel, which is pressed upward into a tunnel ceiling. The bottom of the pan is sloped to a corner where the outlet is located, and tubing conveys the soil water to a collection vessel. An inert nylon wick may be placed over the packing material to reduce the potential for surface tension at the air-water interface.[20] Increasing the surface area of the sampler will make it more likely that the sampler will intercept a sufficient amount of macropore flow.[12] Wick samplers may appear similar to pan types, but an inert wick (usually fiberglass) is incorporated into the design to pull water under tension into the sampler in addition to intercepting macropore flow.[4]

Preparation and installation of percolate samplers

An appropriate and approved safety and health plan must be prepared and used for all laboratory and field activities (see Chapter 6). Prior to emplacement, the sampler body should be cleaned as described for a vacuum sampler. A predetermined number of samplers should be filled with the proposed packing material (if other than soil) and deionized water should be run through the sampler, including sampling lines and collection vessels. A suitable number of equipment blanks should be taken prior to emplacement to ensure that the samplers are adequately cleaned and are capable of obtaining samples that are not contaminated before installation. When the costs of installation are considered, preinstallation quality control samples are prudent.

Cleaned samplers must be carefully dried, and transported to the site properly wrapped in clean plastic. New packing material should be used at the site, taking care not to contaminate it by adding foreign material including solids, liquids, and gases. Disposable plastic gloves should be worn during installation. A space of 5 to 10 mm below the edge of the pan should be left unfilled to allow room to press the pan upwards into the tunnel ceiling.

In accordance with the appropriately approved site-specific health and safety plan and OSHA Regulation 1926.651, all utilities and pipeline companies need to be properly notified. Installation of a percolate soil water sampler begins by excavating a trench, which is then adequately shored against caving in. A bench is then cut into the wall of the trench at the desired sampling depth. A shelter is sometimes constructed in the trench to gain access to the sampler after installation. If a shelter is not desirable, then a pressure-relieved collection sump is installed, with tubing running to ground surface. The ceiling above the bench is carefully scraped with a clean hand trowel to reduce smearing. The sampler is pressed upward into the ceiling and is held in place by fixed bracing, air-filled pillows, or screw jacks. All edges of the sampler must be in firm contact with the soil ceiling or soil water may bypass the sampler.

Sample collection

Depending on the soils and depth of the sampler, a sample may be acquired after a precipitation event or an artificial hydraulic loading. Percolate samplers must be drained after each wetting front has passed to prevent overflows and cross contamination between sampling events. If a shelter was constructed then the sample is manually removed from within the shelter. If a collection sump was installed, then the sample is removed as described for a vacuum soil water sampler. Without check valves, pressure cannot be used to force a sample from a sump, and if a vacuum is required to obtain the sample then volatile organic compounds may be compromised with this type of sampler configuration.

Quality assurance/quality control

Quality assurance/quality control for soil water samplers must be done in accordance with the QA/QC plan that governs soil water sampling at a particular site. At minimum this consists of:

- Adequate precleaning, and proper installation of the sampler
- Adequate documentation of precleaning, installation, and field conditions and procedures used when sampling
- Purging the sampler of native soil water prior to sampling (vacuum samplers only)
- Removing the sample from the sampler within prescribed holding times
- Taking two samples from the same sampler within a relatively short time frame and comparing the analytical results with acceptable variance limits
- Instituting adequate QA/QC procedures typically used for collecting ground water samples (see Chapters 6 and 7)

The B.A.T.® system

A relatively new type of soil water sampler is called the B.A.T.® system. The B.A.T.® has a ceramic tip and is installed in a bore hole with a 2-in.-diameter well casing extending to the ground surface. Precleaned, evacuated 500-mL sample vials are lowered to the tip in a messenger unit, and a hydraulic connection with the tip is established with a double-ended hypodermic needle. The vacuum in the sample vial draws the sample into the vial, and when a sufficient sample volume is obtained the messenger unit with the sample vial is retrieved to the surface.

Summary

Soil water samplers may provide useful information on water chemistry in the vadose zone. The inherent limitations of soil water samplers must be evaluated prior to installation, and proper installation and sampling procedures must be followed to ensure that samples are obtainable and are representative.

References

1. **Anon.,** *Glossary of Geology,* American Geologic Institute, Alexandria, VA 1982.
2. ASTM, *Standard Guide for Pore-Liquid Sampling from the Vadose Zone,* D4996-92, American Society for Testing and Materials, Philadelphia, PA, 1992.
3. Ballestero, T., Herzog, B., Evans, O.D., and Thompson, G., *Ground Water Monitoring,* Lewis Publishers, Chelsea, MI, 1992.
4. Boll, J., Selker, J.S., Nijssen, B.M., Steenhuis, T.S., Van Winkle, J., and Jolles, E., *Water Quality Sampling Under Preferential Flow Conditions,* Lysimeters For Evapotranspiration and Environmental Measurements, ASCE, American Society of Civil Engineers, Washington, D.C., 1992.
5. Bumb, A.C., McKee, C.R., Evans, R.B., and Eccles, L.A., Design of lysimeter leak detector networks for surface impoundments and landfills, *GWMR,* Spring, 1988.
6. Cullen, S.J., Kramer, J.H., Everett, L.G., and Eccles, L.A., Is our ground water monitoring strategy illogical?, *GWMR,* Summer, 1992.

7. Creasey, C.L. and Dreiss, S.J., Porous Cup Samplers: cleaning procedures and potential sample bias from trace element contamination, *Soil Science*, 145(2), 98–101, February, 1988.
8. Davis, S.N. and deWiest, R.J.M., *Hydrogeology*, John Wiley & Sons, New York, 1966.
9. Everett, L.G., Hoylman, E.W., Wilson, L.G., and McMillion, L.G., Constraints and categories of vadose zone monitoring devices, *GWMR*, Winter, 1984.
10. Everett, L.G., McMillion, L.G., and Eccles, L.A., *Suction Lysimeter Operation at Hazardous Waste Sites, Ground Water Contamination,* ASTM Field Methods, American Society for Testing and Materials, Philadelphia, PA, 1988.
11. Hillel, D., *Fundamentals of Soil Physics*, Academic Press, New York, 1980.
12. Jemison, J.M., Jr. and Fox, R.H., Estimation of zero tension pan lysimeter collection efficiency, *Soil Science*, 154(2), 85–95, 1992.
13. McGuire, P.E., Lowery, B., and Helmke, P.A., Potential sampling error: trace metal adsorption on vacuum porous cup samplers, *Soil Sci. Soc. Am. J.*, 56, 74–82, 1992.
14. McGuire, P.E. and Lowery, B., Evaluation of several vacuum solution samplers in sand and silt loam at several water potentials, *GWMR*, Fall, 1992.
15. McGuire, P.E. and Lowery, B., Monitoring drainage solution concentrations and solute flux in unsaturated soil with a porous cup sampler and soil moisture sensors, *Ground Water*, 32(3), 356–362, 1994.
16. Morrison, R.D. and Lowery, B., Effect of cup properties, sampler geometry, and vacuum on the sampling rate of porous cup samplers, *Soil Science*, 149(5), 308–316, 1990.
17. Peters, C.A. and Healy, R.W., The representativeness of pore water samples collected from the unsaturated zone using pressure-vacuum lysimeters, *GWMR*, Spring, 1988.
18. Silkworth, D.R. and Grigal, D.F., Field comparison of soil solution samplers, *Soil Sci. Soc. Am. J.*, 45, 1981.
19. Soil Measurement Systems (1994 sales literature), Tucson, Arizona.
20. Swistock, B.R., Yamona, J.J., Dewalle, D.R., and Sharpe, W.E., Comparison of soil water chemistry and sample size requirements for pan vs. tension lysimeters, *Water, Air Soil Pollution*, 50, 387, 1990.
21. USEPA, EPA Permit Guidance Document on Unsaturated Zone Monitoring for Hazardous Waste Land Treatment Units Section 4: Soil Pore Liquid Monitoring, U.S. Environmental Protection Agency, Washington, D.C., 1985.
22. USEPA, Monitoring in the Vadose Zone, A Review of Technical Elements and Methods, U.S. Environmental Protection Agency, Washington, D.C., 1990.
23. Wildenschild, D., Jensen, K.H., Villholth, K., and Illangasekare, T.H., A laboratory analysis of the effect of macropores on solute transport, *Ground Water*, 32(3), 1994.
24. Wood, W.W., A technique using porous cups for water sampling at any depth in the unsaturated zone, *Water Resources Research*, 9(2).

chapter three

Design and installation of monitoring wells

The best man is like water. Water is good; it benefits all things and does not compete with them. It dwells in lowly places that all disdain. This is why it is so near to Tao.
Lao-tzu, ca. 604–531 B.C.

Locating monitoring points

Prior to the installation of monitoring wells decisions have to be made about where to locate them, what materials should be used, and what construction techniques are appropriate. The spatial configuration of a monitoring system is predicated on many variables including:

- Nature of the wastes (e.g., landfill vs. a spill)
- Soils and geology
- Distribution of contaminants in the vadose zone and aquifer
- Aquifer characteristics including hydraulic conductivity, horizontal and vertical homogeneity, and hydraulic gradient
- Physical and chemical characteristics of the contaminant(s) including concentrations, solubilities, and densities
- Environmental impacts and human health risks
- Where the site is in the regulatory process

As much information as possible about the site must be assimilated before the installation of monitoring wells so that an accurate conceptual model — which includes soil, geologic, hydrogeologic, and geochemical component — can be used to design the environmental monitoring system (EMS).

After compiling and assessing existing information, it may be necessary to further define the soils and geology of the site with test borings prior to installation of the monitoring well and to install piezometers so that the direction of ground water flow is established. Qualitatively, geochemically profiling the site with push-type samplers, performing a soil gas survey, or using geophysical methods prior to well placement should be considered to augment the information derived from borings and piezometers.

Soil borings

As appropriate, an approved safety and health plan must be prepared and used for all laboratory and field activities (see Chapter 6). In accordance with the site-specific safety and health plan and OSHA Regulation 1926.651 all utilities (including gas, electric, phone, water and sewer) and all underground pipeline companies must be adequately notified about when and where drilling or trenching is to occur so the utility or company can mark the locations of their lines.

The locations and depths of soil borings to be done prior to installation of the EMS should be proposed in a formal work plan, reflecting the spatial orientation of the site.[44] Typically, samples from borings are classified and collected continuously for the first 15 ft (where the most rapid lithologic changes usually occur), and then every 5 ft or with each change in lithology. A professional with expertise in hydrogeology (as defined by education and experience) must do the field logging and assist in making changes to the work plan in the field, as required.

Field classification of the borings is typically done using either the ASTM method D 2488 or the USDA nomenclature. The ASTM classification system provides information on compressibility, shear strength, load bearing, and some general color resolution. The USDA system provides information on water movement and the texture and structure of the soils. Using the ASTM field classification method with the USDA colors is generally acceptable.

Soil samples can be collected with hand-operated soil samplers from adequately shored trench walls or in conjunction with drill rigs. Hand-operated devices include screw-type augers, barrel augers, tube-type augers, and hand-held power augers. Soil samplers used in conjunction with drill rigs include split spoon samplers, Shelby (thin wall) tubes, ring-lined barrel samplers, continuous sample tube systems, and piston samplers (Figure 3.1).

The protocol for obtaining soil samples for contaminant analyses is different than formation sampling for lithologic information, which is outlined below. The most commonly used devices for obtaining samples for lithologic information with drill rigs are split spoon samplers or Shelby tubes. In some instances it may be appropriate to obtain both split spoon and Shelby tube samples in the same bore hole if a liner is not used in the split spoon sampler.

A split spoon sampler is a hollow metal tube (most commonly a 2-in. outer diameter and a 1.5-in. inner diameter) which, after sampling, opens along its major axis for sample recovery. For use, the split spoon sampler is attached to a sampling rod and is lowered into the borehole. A 140-lb hammer is dropped on the rod assembly from a distance of 30 in. and the number of blows required for each 6-in. increment of the sample rod is recorded. The sampler is advanced for 18 in., or until there is no observed advance of the sampler within a specified number of blows. The first 6 in. are considered to be the seating drive, and the sum of the next two 6-in. increments is termed the *standard penetration resistance*, or the *N-value* (ASTM method 1586-84).[3]

Figure 3.1 A split spoon sampler opened with a liner (the two pieces on the left) and a Shelby tube sampler (on the right).

Split spoon samplers can be fitted with a plastic or metal retaining basket to prevent noncohesive samples from falling out during retrieval.

A split spoon sample is used for field characterization. Representative, undisturbed section(s) of the core are put into appropriately cleaned, labeled, and sealed moisture-proof jars to be transported to a laboratory for further testing, including grain size distribution. Due to sample disturbances, laboratory permeability tests are not recommended for samples obtained from a split spoon sampler unless the sampler is lined with a close-fitting liner. In relatively homogeneous media, multiple brass tubes with a length of approximately 2 in. each may be used to line a split spoon sampler to provide lithologic resolution and samples for laboratory testing.

Shelby tube samples are appropriate in clays, silts, and fine-grained sands, as in noncohesive soils the sample may not be retained (ASTM D 4700-91).[2] Shelby tube samplers are stainless steel tubes, commonly 30 in. long with a 3-in. diameter. The tube is attached to a rod and is placed in the bottom of the borehole. The tube is pushed into the sample zone by a continuous rapid motion imparted by the drill rig's hydraulic ram, and the length of advance is determined by the resistance and condition of the formation. The length of advance should never exceed 5 to 10 diameters of the tube in sands and 10 to 15 diameters of the tube in clays (ASTM method D 1587-83).[4] The tube may be driven, but this may cause it to (more readily) buckle. If the tube is driven, the sample must be labeled as a driven sample.

Suction may develop between the sampler–soil interface; therefore, a Shelby tube sampler should be twisted prior to extraction or a portion of the sample may remain in the borehole. When the tube is removed from the borehole the bottom 1 in. of sample is removed for field classification. The tubes are then sealed with end caps, properly labeled, and shipped to the laboratory in an upright position. Laboratory-derived vertical permeability data may be obtained from a Shelby tube sample.

Jar headspace measurements may need to be taken with an appropriately calibrated organic vapor monitor (OVM) equipped with a photoionization detector (PID) or flame ionization detector (FID). Taking a headspace sample can be accomplished by placing the soil in a plastic bag and shaking the bag, or by half-filling a 500-mL jar with approximately 250 mL of soil. If a jar is used, then aluminum foil is put over the top of the container and the container is vigorously shaken. After 10 min the bag or jar is vigorously shaken again. The bag or foil seal is then punctured, and the instrument's sampling probe is quickly inserted to approximately one-half of the head space depth, using care not to allow water or soil to be drawn into the probe tip. Maximum response should occur within 2 to 5 s and the maximum response should be recorded on the field boring log.

An explosivity meter (if needed) is usually employed at appropriate intervals in the top of the borehole or the top of the hollow stem auger. Action limits for elevated gas concentrations should be specified in the site-specific safety and health plan.

Investigative derived wastes (IDW) from site investigations may pose a risk to human health and the environment. IDW may include drilling muds, cuttings, soil and other samples and well development and purge water. A management plan for IDW should consider the following in making a best professional judgment for the disposition of IDW:

- The contaminants, their concentrations, and total volume of IDW
- Media potentially affected (e.g., ground water, soil)
- Location of the nearest population(s) and the likelihood and/or degree of site access
- Potential exposures to workers
- Potential for environmental impact

"Until there is positive evidence (records, test results, other knowledge of waste properties) that the IDW is a RCRA hazardous waste, site managers should manage it in a protective manner (but not necessarily) in accordance with Subtitle C requirements."[45] If IDW is determined to be a RCRA hazardous waste, then land disposal restrictions may apply.

Field classification of soils and geologic samples is undertaken to identify formation changes, to facilitate changes to the work plan, and to ensure that appropriate samples are sent to the laboratory for further analysis. The field log must include general information such as the project name, hole number, date started and completed, geologist's and driller's names, hole location (detailed, scale map), rig type and drilling method, sampling equipment used, classification scheme, and description of salient features in proximity to the borehole. The borehole log must also include the following field information:

- Observations on blow counts, advance rate, and sample recovery
- Texture (percent sand, silt, and clay)
- Grain size, sorting, and degree of grain rounding (angular to spheroid)
- Vertical extent of soil zones
- Primary and secondary voids
- Structural features such as bedding planes, cross beds, graded features, lineations, solution channels, discontinuities, bioturbations, and respective orientations
- Moisture content (wet, moist, dry)
- Color (refer to a Munsell chart) and presence of discoloration or mottling
- Depending on the site, OVM or Geiger counter readings of the samples and explosivity meter readings from the borehole
- Depth of borehole and depth of each water bearing unit
- Depth, location, and identification of any contaminants encountered
- Surveyed horizontal location and elevation of the top of the borehole
- Other comments or observations, such as deviations from the drilling plan or method

The professional in charge of doing the borings must be constantly reassessing the preliminary model and revising the field work accordingly. This could entail relocating borings or doing more or deeper borings than originally anticipated. Drillers usually charge a mobilization fee, so doing the additional borings at the same time as the originally planned borings can save time and money.

Interstate transportation of soil samples is subject to regulations established by the US. Department of Transportation (DOT); Agriculture, Animal, Plant, and Health Service; plant protection and quarantine programs; and possibly other federal, state, and local agencies.[9] The DOT rules that govern the transport of samples are found in *49 CFR 171-179*.

After soil samples are transported to the laboratory, analyses include producing grain size distribution curves (using sieves and/or hydrometers) and determining porosity and vertical permeability. In addition, determining clay minerology may assist in assessing aquifer continuity. All borings that are not subsequently used for monitoring point construction must be properly sealed with bentonite or neat cement grout to minimize the potential for future pollutant movement along the borehole.

Drilling for monitoring point installation

Methods for drilling monitoring wells include hand augering, driving wells, jet percussion drilling, using solid flight and hollow stem augers, direct mud rotary, air rotary or air percussion, air rotary or percussion with a casing driver, dual-wall reverse circulation, cable tool, bucket augers, reverse circulation rotary, and sonic drilling. The drilling method chosen must:

- Facilitate formation sampling (soil, unconsolidated material, rock)
- Introduce a minimum amount of drilling fluids (including air)
- Keep wall smearing and sealing at a minimum
- Keep the bore hole open for the installation of the well
- Allow a minimum of a 4-in. annular space around the well casing and screen for placement of the filter pack and annular seal
- Avoid contamination or cross-contamination of ground water and aquifer materials
- Reduce the volume of potentially contaminated cuttings

When drilling for core sampling, piezometer, or monitoring well installation, an approved soil sampling procedure should be followed. At minimum, soil samples should be taken for field classification continuously for the first 15 ft and then every 5 ft or with each change in lithology. Continuous logging may be required in some situations. A drilling plan in which a monitoring well is located immediately adjacent to a previously completed borehole can obviate the need for coring and logging the actual monitoring well if the stratigraphy is analogous in the previously drilled and logged borehole to that in the proposed monitoring well.

The most commonly used drilling tool for taking borings and for installation of monitoring wells in unconsolidated materials is the hollow stem, continuous flight auger (Figure 3.2). Each flight (section) is 5 ft in length, is hollow, and has large external threads. The first flight has a cutting head, and as the auger is rotated and driven into the soil the auger "screws" into the ground, moving the cuttings to the surface, and provides a temporary casing. More flights are added as required. The hollow flights provide access for formation sampling as drilling progresses and also allow for installing the screen and casing string, filter pack, and annular seal for the monitoring well, as the hollow flights act as a temporary casing. Some drill rigs capable of drilling with hollow stem augers can also be used for mud rotary drilling.

The use of a pilot bit inside of the hollow stem auger is recommended to reduce the potential for formation material to be drawn up into the hollow stem, making for unrepresentative boring cores. A pilot bit may also reduce the problems associated with heaving sands. If the depth of the well or the

Figure 3.2 A hollow stem auger drill rig.

nature of the formation materials precludes the use of a hollow stem auger rig, then an alternative method of drilling must be selected.

When the water table is encountered in unconsolidated formations, mud (a combination of bentonite clay and water) is sometimes used with a hollow stem to keep the formation from flowing up between the inside of the auger and the pilot bit, causing them to lock up together. The use of a smaller diameter tri-cone bit allows mud to be pumped into the formation. Mud keeps the formation from flowing up into the hollow stem and locking up the tools and also reduces the potential that the next soil sample contains material that was drawn up into the auger, which would make it unrepresentative. The use of drilling mud may be of concern due to chemical and hydraulic alterations of the surrounding media. Figure 3.3 shows various types of drill bits.

Other methods commonly selected for taking borings and for monitoring well installation include the use of a cable tool, or an air or mud rotary rig. A cable tool drill rig utilizes a heavy weight (drill bit) which is repeatedly raised and dropped to pound a hole into the ground (Figure 3.4).

Figure 3.3 From left to right, a hollow stem auger bit, two types of roller cone drill bits (one with carbide buttons for air rotary drilling), and a drag bit. (Photograph courtesy of WTD Drilling, Inc.)

Cable tool drilling is relatively slow. Cuttings must be periodically removed from the borehole with a special bailer by adding water, if not naturally present. If the formation is unconsolidated then the casing needs to be driven simultaneously with drill bit advancement. Cable tool drill rigs are sometimes selected for installation of recovery wells, as they may cause less smearing of the formation in comparison to other drilling methods (Figure 3.5).

An air or mud rotary drill rig imparts a rotary action and a downward pressure to drill rods to which a drill bit is attached (drag or roller cone) to auger through the soil (Figure 3.6). Cuttings are brought to the surface and the borehole remains open by using high-pressure air (air rotary), pressurized clean water, or a variably viscous mixture of water and bentonite clay called mud (mud rotary) down through the drill rods, out through the bit, and back to the surface through the borehole. The air, water, or mud also serves to lubricate and cool the bit, prevent inflow from aquifer materials, and reduces the potential for cross contamination between aquifers.

When using an air rotary or air percussion drill rig, adequate air filters must be used on the compressor to prevent lubricating oils from invading the formation (Figure 3.7). Drilling foams that contain organic chemicals may assist in bringing cuttings to the surface but should be avoided to prevent potentially altering the chemistry of the aquifer. A combination of air rotary or air percussion hammer drilling while simultaneously driving a casing as

Figure 3.4 A capable tool drill rig.

the borehole is advanced may be a viable drilling technique, but obtaining representative formation samples as drilling progresses entails breaking down the drill string every time a sample is required, and this can be time-consuming.

The drilling fluid used for mud rotary drilling can invade the formation, making it difficult to remove from the filter pack, may reduce ground water flow, and may potentially alter the ground water chemistry.[27] The cuttings produced by monitoring well drilling may be contaminated (an investigative derived waste) and, if so, must be properly disposed of. Wells drilled into dolomite bedrock with a rotary-wash method (using water as the drilling fluid) averaged an order of magnitude less transmissivity than wells drilled into the same dolomite using air rotary methods.[40]

Rotosonic drilling is fast and provides a continuous core sample with limited cuttings.[20] A hydraulically activated sonic head provides high-frequency sinusoidal waves (frequencies of 150 Hz) which are imparted into the

Figure 3.5 A (large) cable tool drill bit (pencil for scale).

drill string, and from there into the drill bit (Figure 3.8). A drill bit is attached to the drill string, and the string is oscillated, rotated, and hydraulically pressed downward. One casing size (4.5 in. in diameter) is used for formation sampling and a larger size (6 in. in diameter) is used for well installation.

Rotosonic drilling may be more expensive than using a hollow stem or mud rotary, but the excellent lithologic resolution, speed, and limited cuttings production may offset the additional expenses. Heaving sands may be a problem when using a rotosonic drill rig.

Cleaning drilling equipment

Drilling and sampling equipment should be cleaned between boreholes, and the degree of decontamination between holes depends on the site. The main types of drilling decontamination equipment are steam cleaners, low- and high-pressure hot water washers, and combination units. Steam cleaning

Figure 3.6 An air rotary drill rig with a casing driver.

units typically operate at more than 350°F and pressure washers operate at temperatures from between 120 to 230°F at pressures of between 500 to 3,000 psi.[26] Steam cleaners can quickly take off difficult to remove contaminants with less water than a hot water washer, but may form a mist of water droplets and volatile contaminants. Some types of environmental drilling, such as at radioactively contaminated sites, may require a thorough cleaning of the rig between boreholes at a decontamination station (a bermed pad) where rinse water is collected, tested, and properly disposed of (see Figure 3.9).

A specific cleaning plan should be part of the drilling plan. A decontamination report should be submitted with the well logs, specifying date, time, location, equipment involved, decontamination location, individuals who performed the decontamination, decontamination procedures, source of materials used for the decontamination, handling of the rinse fluids, and any other analyses performed.[7]

Figure 3.7 An air percussion drill bit.

At minimum, the drill rig should be visibly free of hydraulic fluid, grease, oils, or fuel. This is accomplished by a thorough steam cleaning or a pressure washing of the rig prior to the commencement of drilling at a site. Care should be exercised when washing the drill rig to minimize deterioration of gauges, hoses, wiring, seals, etc. The more sensitive components of a drill rig can be washed with a stiff brush and a nonphosphate inorganic detergent. Auger flights (and sometimes drill rods) should be steam cleaned or pressure washed between holes. Split spoon samplers may be scrubbed with a stiff brush and a nonphosphate detergent solution between sampling intervals. Only chemically inert threading lubricants (such as Teflon™-based) should be used on drill rod connections.

Piezometer installation

The direction of ground water flow should be ascertained prior to the installation of permanent monitoring wells. Accurate, longer-term information on the direction of ground water flow is provided by the installation of piezometers. A piezometer is a device (similar to a monitoring well) which is used to obtain water level measurements, but not ground water samples. Ideally a piezometer has a very small diameter so it can respond quickly to changes in the potentiometric surface. A typical piezometer is constructed from 1.5-in.-diameter PVC casing to facilitate using water level measuring devices including continuous data loggers with pressure transducers. A piezometer usually has a screen length of 2 to 5 ft.

Figure 3.8 The sonic head of a rotosonic drilling rig.

A piezometer may be installed similarly to a monitoring well or can be a driven well point. A drilled piezometer must be installed with a tight annular seal so that ground water does not bypass the seal and flow up or down the side of the casing. A permeable annular seal would provide erroneous water levels in the piezometer and possibly allow contaminants to move deeper into the aquifer. When piezometers are drilled, the new boring logs obtained supplement the previous boring information.

Vertical ground water flow information may be obtained by installing a deeper piezometer immediately adjacent (nested) to a shallower piezometer. These multiple, immediately adjacent installations are called *nested piezometers*, and they are either installed in the same borehole or in immediately adjacent boreholes.

All borings and monitoring points including piezometers and monitoring wells must be installed in accordance with local well codes. After installation, piezometers must be developed and accurately surveyed vertically to the nearest 1/100th of a foot by using permanent site benchmark(s). Subsequent to installation, development, and surveying, water levels in the piezometers should be recorded over as much time as can be afforded, as water levels can change substantially seasonally and from year to year.

Monitoring well installation

After completion of the preliminary field work including borings, piezometer installation, and taking water level measurements, the conceptual model

Figure 3.9 High-pressure washing of a hollow stem flights.

of the hydrogeology is further refined. Geologic cross sections of the site, potentiometric surface maps generated from water levels taken from the piezometers, evaluation of horizontal and vertical hydraulic conductivities and ground water flow rates, and information derived from any other investigative work undertaken at the site (possibly including seismic profiling, push-type samplers, etc.) are used to add to the conceptual model. The refined model is the basis for selecting the construction materials and techniques, number, horizontal locations, and the vertical screened intervals of permanent monitoring wells in an EMS.

The most basic EMS for ground water monitoring consists of three water table monitoring wells, one located hydraulically upgradient and two downgradient of the site. This basic EMS configuration allows for a comparison of unimpacted ground water chemistry (the upgradient well) to the two downgradient wells and allows an elementary planar analysis of ground water flow, since three points in space (the three water levels in the three wells) define a plane. Three monitoring wells may be insufficient to define the hydraulics and geochemistry of a site and more wells, including nested wells which are in line with the direction of ground water flow, may be required.

Errors can be introduced in VOC concentrations due to purging partially penetrating wells (screens which bridge separate aquifers/aquacludes), and for comparison purposes of hydraulics and chemistry monitoring wells need to have similar screened intervals and depths.[39]

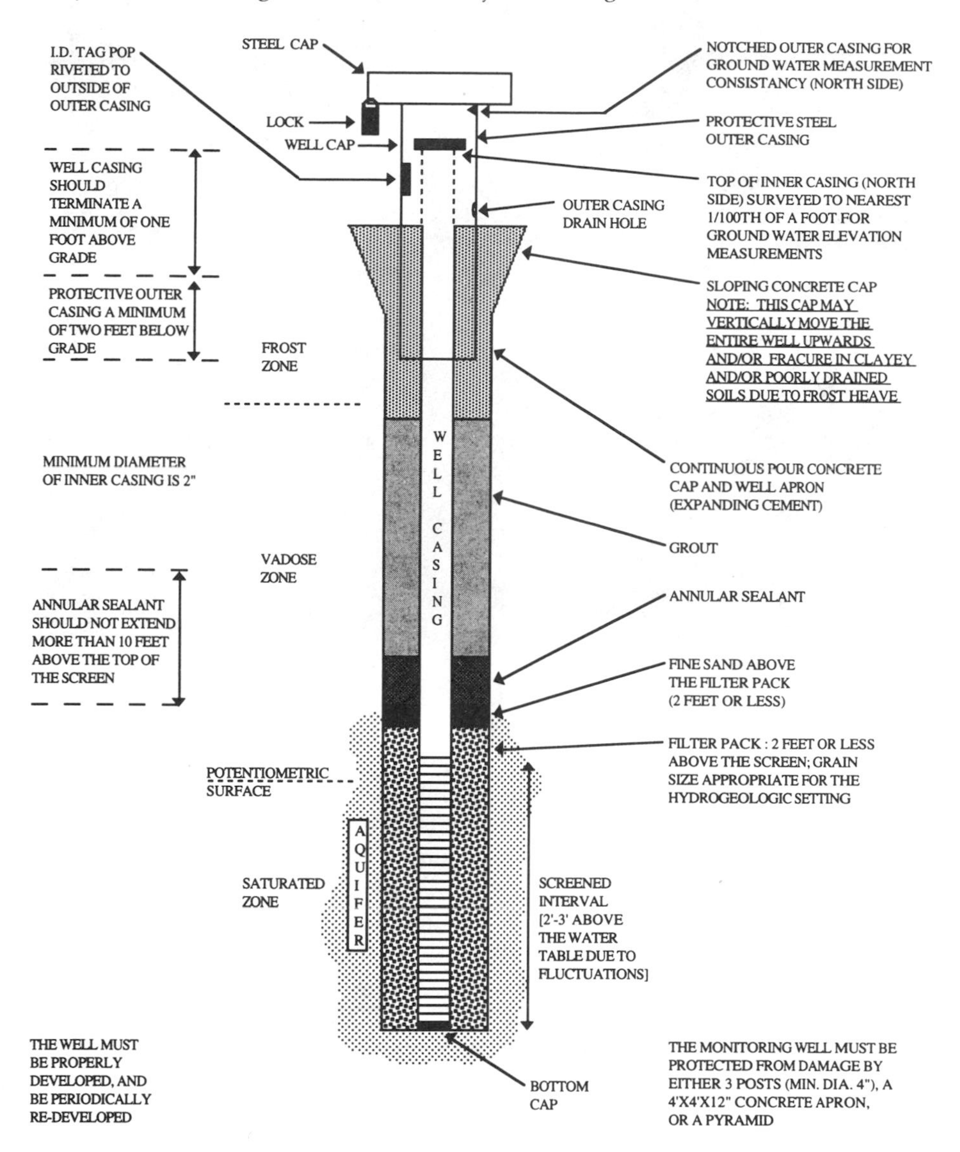

Figure 3.10 A construction diagram of a typical monitoring well.

A typical monitoring well has a minimum diameter of 2 in. and consists of a screen with a sump, filter pack, annular seal, riser pipe, surface seal, and locked outer casing (Figure 3.10). The purpose of a monitoring well is to provide information on ground water levels and chemistry. The materials used to construct a monitoring well may leach or adsorb chemical constituents and alter natural ground water flow. A basic goal for the design and installation of monitoring wells is to reduce alterations of flow and chemistry as much as possible to obtain representative water levels and samples.

Figure 3.11 A monitoring well screen (keys shown for scale).

Monitoring well screen (intake) design is becoming more standardized, and although other types of screen designs are available, a continuous-wrap, wire-wound screen for PVC (polyvinylchloride) or stainless steel screens or factory slotted screens for PVC wells are usually selected based on availability and performance (Figure 3.11). Manual slotting of screens in the field should not be permitted under any circumstances.

A 10-ft-long screen is usually selected if it is to straddle the water table, or a 5-ft-long screen if below the water table. If the goal of monitoring is to intercept floating free product (e.g., gasoline), a longer screen may be preferred to ensure that a sufficient length of the screen will always be above the water table.

A monitoring well screen provides an integration of flow rates and contaminant concentrations along its length. This integration can result in a contaminant concentration that may be as little as 28% of the maximum concentration obtained in a multiport sampler.[17] Using multiple, smaller intakes along the screened interval may provide better flow and geochemical resolution but correspondingly increases analytical costs (see Chapter 5).

Resistance of materials to chemical adsorption and leaching

Materials such as Teflon™ (PTFE) and fiberglass-reinforced epoxy resin are available for use in monitoring wells, but the most commonly used screen materials are rigid PVC and stainless steel (SS) 304 or 316. SS 304 costs less than SS 316, but SS 316 performs better under reducing ground water conditions and when the ground water has sulfur-containing species. The most commonly used riser materials are PVC and low-carbon steel.

To reduce materials costs, a stainless steel screen may be attached to a 10-ft length of stainless steel riser; black pipe (low-carbon steel) is attached to the stainless steel riser that extends to the ground surface. Unless a large potential exists for corrosion, a dielectric coupling is usually not warranted to prevent galvanic action from occurring between a stainless steel screen attached to a low-carbon steel riser pipe, as the metals have similar relative potentials.[15]

There are several factors to consider when selecting screen and riser materials including the importance of the material being chemically inert (not leaching or adsorbing chemical constituents), the material's resistance to chemical attack, screen and casing strengths, and costs. Several studies have been undertaken to assess leaching or adsorbance of chemical constituents with screens and casing materials, and these are summarized as follows:

- PVC casing material adsorbs fewer organic contaminants than PTFE (Teflon™), and SS 304 and SS 316 casing materials do not adsorb organic compounds.[37]
- PVC casing material adsorbs organic compounds at such slow rates that if the well is purged of stagnant water within 12 hours of sampling, the sample will not be significantly affected by sorption processes. PTFE, however, showed significant uptake of at least one organic compound within 3 hours.[19]
- Concentrations of methylene chloride below 3340 mg/L (0.2 activity solution) will not cause softening or serious swelling of rigid PVC. PVC exposed to 16,700 mg/L (1.0 activity solution) of methylene chloride softened within the first 4 days of exposure to the test solution.[36] At ppm levels, the rate of diffusion of organic contaminants within PVC and PTFE is independant of concentration.[46]
- *Cleaners or glues must not be used for PVC wells monitoring VOCs,* as cleaners and glues will leach significant amounts (in some cases mg/L) of numerous organic compounds, notably methyl ethyl ketone and tetrahydrofuran, for long periods of time after well construction.[14] Cleaners and glues can also mask the presence of other priority pollutants.[31]
- Surface characteristics of PVC, PTFE, and SS 304 and SS 316 exhibited no changes when exposed to mg/L concentrations of several organic compounds after 6 months.[42]
- The rate of uptake of organic compounds onto PVC casing material increases with decreasing water solubility and with increasing octanol-water partition coefficient.[19]
- Stainless steels impose a chemical signature on ground water samples and are therefore not suitable where trace metal determinations are planned.[22]

- PTFE casing material is inert with respect to trace metals.[24]
- PVC casing materials leach metals, but at low concentrations that decrease with time. SS 304 and SS 316 leach significant concentrations of metals that increase with time.[23] SS 304 can leach chromium, nickel, molybdenum, iron, titanium, cobalt, and tungsten.[31]
- Stainless steel corrodes in oxidizing environments, and a potentially reactive hydrous iron oxide precipitate may develop on SS 316 at low pHs, causing a random source of error.[22]
- Leaching of iron, cadmium, chromium, and manganese have been observed in wells constructed of stainless steel, even after purging the well of stagnant water.[11,29]
- Under dynamic flow conditions (water flushed through screens) both SS 304 and SS 316 leached copper and nickel in excess of five times their respective minimum detection limits; iron was adsorbed by SS 304 and more so by SS 316. It was demonstrated that PVC and Teflon™ screens adsorb lead and cadmium, but SS 304 and SS 316 sorption of lead and cadmium was significantly greater than that by the polymeric casings. Air bubbles initially trapped on the polymeric screens was thought to reduce iron concentrations, and active corrosion on the stainless steel screens was thought to provide a mechanism (galvanic corrosion) for the loss of ferrous iron from solution. No significant sorption of trichloroethylene with initial concentrations of between 0.5 to 2.0 mg/L was observed for SS 304, SS 316, Teflon™, or PVC screens under dynamic flow conditions. According to Hewitt,[25] "Only PVC and PTFE, which showed no influence or diminished influences in comparison with stainless steel screens, should be recommended for construction of wells intended for monitoring metals in ground water." Stainless steel would probably not be considered an appropriate material in corrosive ground water or where determinations of trace metal concentrations are of primary concern. Likewise, PVC probably would not be considered an appropriate material in situations where solvents of moderate to high concentrations might dissolve the PVC material.[47]

With respect to being chemically inert, steam-cleaned National Sanitation Foundation-approved PVC casing with joints sealed with PTFE tape or an inert "O" ring appears to be a reasonable compromise for monitoring lower levels or the early detection of organic contaminants simultaneously with monitoring trace metals, if the well is properly constructed. PVC may be a good choice because SS 304 and SS 316 may bias trace metal concentrations in ground water samples; PVC has only limited adsorption potential with respect to lower concentrations of organic compounds, and PVC costs considerably less than stainless steel or PTFE. Some PVC pipe manufactured in foreign countries may contain significant quantities of lead and so may not be suitable for monitoring well installation.

Resistance to chemical attack is a concern for both PVC and stainless steel casing materials. PVC casing may deteriorate (solvate) in the presence of low molecular weight ketones, amines, aldehydes, and chlorinated alkenes and alkanes.[1] At sites where elevated concentrations of the above-listed LNAPLs (light nonaqueous phase liquids, i.e., "floaters") and DNAPLs (dense nonaqueous phase liquids, i.e., "sinkers") are present, PVC pipe should not be used because floaters and sinkers may become concentrated in the water column, and this may accelerate PVC solvation. If any of the above-listed soluble contamination concentrations are approaching that of pure product or if any of the above contaminants are concentrated in the soils that are in direct contact with the screen or riser pipe, PVC materials should not be used in order to preclude solvation and subsequent screen or casing failure.

If any of the following conditions exist in the ground water, then the possibility of corrosion of stainless steel must be considered, and combinations of these conditions make corrosion of stainless steel screen or casing more likely:[15]

- Dissolved oxygen concentrations are greater than 2 mg/L
- The pH of the ground water is less than 7
- Total dissolved solids are greater than 1000 mg/L
- Hydrogen sulfide concentrations are greater than 1 mg/L
- Carbon dioxide concentrations are greater than 50 mg/L
- Chloride concentrations are greater than 500 mg/L

Strength of materials used in monitoring well construction

Materials used in the construction of monitoring wells must have adequate tensile, compressive, and collapse strengths. Tensile strength is the greatest longitudinal (pulling lengthwise) stress the material can sustain before it pulls apart. Compressive strength is the maximum compressive stress (squeezing lengthwise) that can be applied before the material deforms. Collapse strength is perpendicular to the major axis, and is defined as the ability of the material to resist collapse to any external force during and after installation. Collapse strength is proportional to the cube of the wall thickness.

After the borehole is drilled, the screen and casing sections are joined together and lowered into the hole, section by section. The sections are joined either by welding (for metal screen and riser pipe) or using threaded couplings (for metal or PVC screen and riser pipe). Sufficient tensile strength is required for the casing sections (as they hang over the hole) so that the topmost sections do not pull apart from the rest of the already-joined casing string. Depending on how casings are joined, the most likely point of failure is at the casing joints. Dividing the tensile strength by the linear weight of the casing will provide the maximum depth that a dry string of casing can be suspended. If PVC casing is used and the borehole is partially filled with water, a buoyant force will partially offset tensile stresses.

Threading may be either flush threading, where the two coupled pieces form a flush joint, or a pipe thread, where an external coupling is required to join the two sections. Flush threading may be preferable because an external coupling will limit the annular space between the casing and bore hole wall, and annular space is at a premium when emplacing the filter pack and grout. Flush threads must be machined or molded onto the casing at the factory to ensure uniformity of the threading. Flush threads may reduce the tensile strength of the casing string by about 30% as compared to a solvent-cemented joint.[1] Piezometer, but not monitoring well construction, could include using glued joints.

The use of 2-in. nominal diameter schedule 80 pipe (wall thickness of 0.218 in.) instead of schedule 40 pipe (wall thickness of 0.154 in.) will allow much stronger threads to be incorporated into the casing. The inside diameter of schedule 80 pipe is 1.939 in.; schedule 40 pipe, 2.067 in. If sampling equipment requires a minimum of a 2-in. inside diameter, then using schedule 80 pipe may not be possible without going to a 4-in. nominal diameter casing.

Manufactured pipe threads on 2-in. nominal diameter schedule 40 PVC flush joint threading with ASTM standard F-480 of 2 threads per inch of casing has a tensile strength of 2100 lb.[30] The average weight of 2-in. nominal diameter schedule 40 PVC pipe is 0.8 lb/ft of length. The maximum hanging weight recommended by the manufacturer for 2-in. schedule 40 PVC pipe is 525 lb. Based on tensile strength alone, the maximum depth to which a 2-in. nominal diameter schedule 40 PVC well could then be installed without an offsetting buoyant force is 656 ft.

Flush-threaded casing joints should never be overtightened. The joints of monitoring well casing sections must be watertight and the use of Teflon™ tape on the threads or chemically inert O-rings is required. If the fluid (hydrostatic) pressure of the annular seal (grout) is greater than the collapse strength of the casing material, the casing will fail prior to grout hardening. The hydrostatic pressure can be calculated by the following formula:

$$\text{Hydrostatic Pressure (psi)} = [\text{Height of Fluid Column (ft)} \times \text{Fluid Weight (lb/gal)} \times 0.052]$$

The collapse strength for a 2-in. nominal diameter schedule 40 PVC riser pipe is 147 psi.[30] Neat cement grout weighs approximately 15 lb/gal. Calculating from the above formula, the maximum depth before collapse would occur would be 188 ft. If an alternative design using stronger casing materials or less dense grout cannot be accommodated, then adding the grout in stages and letting it harden before the next addition has been suggested.[21] The collapse strength for steel casing is much greater than for PVC casing.

The weight of the casing string should not be allowed to rest on the screen at the bottom of the hole (the casing must be kept suspended over the borehole until the grout sets up), or the compressive force may cause the screen or joints to fail structurally.

Other construction considerations

For deeper wells, the use of a casing centralizer at the bottom of the well screen may be warranted to ensure that the filter pack and the annular seal is evenly distributed in the borehole, and that no thin spots of the annular seal are produced along the side of the well. Care must be exercised when using centralizers anywhere else other than at the bottom of the screen to ensure uniform filter pack and annular seal placement.

A monitoring well must be installed straight and not out of plumb (crooked). A deviation survey should be considered when the hydraulic gradient is so small that water levels in out-of-plumb wells will produce an inaccurate determination of ground water flow.[44] Sampling equipment may also get stuck in out-of-plumb wells. The USEPA has recommended that the maximum deviation should be 1° or less for every 50 ft of depth when using drift indicators. Electronic inclinometers or plumb bobs suspended above the hole, in conjunction with drift templates, are used to determine plumbness of wells.[15] If the well is determined to be out of plumb, then the deviation survey must determine the magnitude and direction of deviation, in order to correct ground water elevations.

Selecting a filter pack and screen size

Screen slot size and the filter pack for monitoring wells are selected based on the D-70 of the formation material to avoid excessive turbidity in ground water samples. The D-70 is taken from the laboratory-derived grain size distribution curve of the finest portion of the geologic media where the screen is to be installed and is equivalent to the sieve size that retains 70% (or passes 30%) of the formation material. The uniformity coefficient is the ratio of the sieve size that retains 40% (or passes 60%) to the effective size.[1]

The filter pack for a monitoring well must be of a uniform, well-sorted grain size and based on the cumulative percent retained of the formation material as determined from a grain size distribution curve.[1] The screen slot size should only be determined only after the filter pack has been specified and should hold back between 85 to 100% of the filter pack material. As appropriate, sieve analyses may be undertaken in the field to facilitate well installation.

Numerous screen slot sizes are commercially available for varying geologic media. If the slot size as determined from a grain size analysis does not exactly match sizes commercially available, the use of the nearest smaller slot size is advised. The screen slot size must be selected to exclude, at minimum, 85% of the filter pack material. The inner diameter of the screen should not be less than the casing diameter to allow the passage of sampling equipment. Special "environmental screens" may be necessary to ensure that the inner diameter in the screen is not smaller than the inner diameter of the casing.

A filter pack may not be required in naturally developed wells or when the well is completed in open bedrock. Naturally developed wells, where the formation materials are allowed to naturally collapse in on the borehole, are only recommended when the natural formation materials are relatively coarse-grained, permeable, and of uniform grain size.[1] The USEPA[44] has recommended that the filter pack grain size should be 3 to 5 times the average grain size (50% retained) of the formation. When the grain size is plotted the gradation should be smooth and gradual.

ASTM D 5092 recommends that the filter pack have a 30% finer (d-30) grain size that is 4 to 10 times greater than the 30% finer (d-30) size of the formation material. A number between 4–6 is recommended if the formation material is fine and uniform and between 6–10 if the formation material includes silt-sized particles and has a highly nonuniform gradation. Filter pack materials must be chemically inert and consist of well-rounded grains of uniform size because the well screen slots have uniform openings. The ideal uniformity coefficient should be 1, and should not exceed 2.5.[6]

Keeping monitoring well construction materials clean

Bags of filter pack sand should be kept dry and separate from potential sources of contamination such as oils, gasoline, hydraulic fluids, and engine exhaust. Clean plastic sheets should be placed under and over bags of filter pack sand during storage and transport until ready for use. Other well construction materials, including the screen and riser materials, must be kept away from potential sources of contamination such as gasoline-powered compressors, oils, hydraulic fluids, etc., during delivery to the site.

The screen must be either certified as factory cleaned and delivered to the drill site in a plastic bag or else adequately steam cleaned or high-pressure water washed in the field immediately prior to installation. Each section of riser must be either steam cleaned or high-pressure washed immediately prior to installation. If a driller needs to hold the screen during installation, the use of disposable plastic gloves should be required. Water used in the cleaning and drilling processes should be from an acceptable (potable) source of known chemistry. Collecting the last rinse water for analysis may be required to prove the efficacy of the cleaning protocol.

Construction of the well

The following is a discussion of how a typical monitoring well should be installed. In some monitoring situations different designs may be required. Installation of monitoring wells needs to be undertaken in accordance with the approved installation plan, safety and health plan, and local well codes. If the presence of explosive or toxic levels of gases is possible when the casing is being installed, then an explosivity meter and an OVM should be employed, and extreme care should be used to avoid introducing a spark (or using a cutting torch).

Prior to emplacing the filter pack and the annular seal, the volume displaced by the casing, filter pack, and grout should be calculated, measured during emplacement, and recorded. A significant disparity between the calculated and actual volume of materials used in the well construction must be accounted for, and in extreme cases the well may have to be immediately decommissioned. When a monitoring well is constructed through more than one aquifer, multiple casings (each sealed off from the one above) may be required to reduce the potential for a hydraulic connection between aquifers.

The previously selected screen and casing are joined section by section as the string is lowered into the borehole. A significant buoyant force may be encountered if a PVC screen and riser is used, which can be counteracted by ballasting with either clean water, formation water, or by using the hydraulic ram on the drill rig.[6]

Between 2 and 10% of the filter pack is usually emplaced in the bottom of the borehole. To preclude bridging (voids), the filter pack may be emplaced by using a hopper (large funnel) and a decontaminated 1.25-in.-diameter tremie pipe, adding clean water to the sand as necessary to keep the sand moving downward in the tremie. The use of a weighted measuring line must be used when emplacing the filter pack and the annular seal to ensure complete coverage. The tremie should be rotated in the hole while emplacing the filter pack and grout to ensure a uniform coverage.

The filter pack is usually extended 2 ft above the top of the screen, unless the presence of another aquifer directly above the screen precludes this. As the filter pack, annular seal, and grout are being emplaced and progress is being gauged by using a tape, the drill flights or temporary casing is simultaneously pulled upward in the borehole and removed from the drill string.

The annular seal

If bentonite (sodium montmorillonite, a clay derived from volcanic ash) is used as the annular sealant, it may be added as dry pellets or chips or as a slurry. If the area above the screen is not saturated, then bentonite slurry may be more appropriate than pellets or chips. If bentonite slurry is used, then a 2-ft-thick layer of fine sand should be placed on top of the filter pack to reduce the potential that subsequent addition of the bentonite seal will disturb the filter pack.

A bentonite slurry is prepared by mixing 15 lb of dry bentonite powder with 7 gal of water to produce 1 ft^3 of bentonite slurry, which should be no less than 30% of solids by weight.[33] If bentonite is added dry (pellets or chips), sufficient potable water must be present in the formation or added from the surface to allow for complete hydration. Upon the addition of clean water, bentonite swells to 10 to 15 times its dry volume.

Sufficient time should be allowed to elapse (as specified by the manufacturer) after emplacement of the bentonite and subsequent well construction activities to allow for hydration and expansion. This time period has been estimated to be 1 to 4 hours, depending on field conditions and the form of bentonite used.[34] Allowance of sufficient time for the pellets or chips to hydrate and expand needs to be addressed in the drill plan.

Several possible reasons for annular seal failure, which allows downward migration of water and/or contaminants, include swelling and shrinkage during setting or curing, poor bonding between the sealing material and the casing, and improper mixing or poor emplacement during construction.

Selection of an annular sealant should be based on the position of the static water level and ground water chemistry. If the water to be mixed with bentonite has a total dissolved solids concentration of 500 mg/L or greater or has elevated chloride concentrations,[1] or if alchohols, ketones, and other polar organic solvents are present in the ground water,[18] bentonite may not properly hydrate with the formation water to provide an effective seal.

The use of bentonite or neat cement may alter the chemistry of ground water samples. This may be caused by improper placement of the grout (intrusion into the filter pack) or by choosing the wrong form of bentonite (e.g., pouring powdered bentonite through standing water instead of using pellets or chips which fall to the bottom of the water and then hydrate and expand). Bentonite may take up or release cations; the combination of bentonite and neat cement may raise the pH.

Grouting the well

After the fine sand and bentonite pellets, chips, or slurry are placed on top of the filter pack, grout must be used to fill the annular space between the casing and the bore hole surface. Grouting is the process of mixing and injecting neat cement, bentonite, or neat cement with bentonite into the annular space between the casing and the drilled hole. Grouting is undertaken to provide an effective hydraulic seal, reduce downward migration of contaminants, and increase the strength of the well.

Neat cement grout is Portland cement mixed in a ratio of 1 bag (94 lb) of cement to 6 to 7 gal of clean water.[6] Volumetric shrinkage of grout mixed in this proportion is approximately 17%. Sometimes, due to mixing difficulties, flash setting, and heat of hydration, additional water is added which may unacceptably increase shrinking and cracking of the grout.

There are five types of cement mixes that may be used to make grout as specified in ASTM C 150, and they are

- Type I: General purpose cement, no special properties
- Type II: Moderate to high sulfate resistance, slow hardening, and low heat of hydration

- Type III: High early strength, high heat of hydration, reduces curing time in cold environments
- Type IV: Produces low heat of hydration, develops strength at slower rates
- Type V: High sulfate resistance

If neat cement is chosen for the grout, then in most situations Type I cement is used. The use of quick setting cements containing additives is not recommended. The setting time for Type I cement is usually from 48 to 72 hours, depending on water content. Neat cement grout can compromise the structural integrity of PVC casing due to the heat of hydration produced while the grout is setting up. This is especially true when a collapse of the formation material creates a large void where more grout is placed. Neat cement may shrink and crack when it dries, but cracking appears to be manifested for only a limited distance.[16] Cracking appears to be most pronounced along the grout/casing interface.

Adding bentonite, sand, or diatomaceous earth to the grout reduces the heat of hydration, and a mixture of neat cement and a small percentage (2 to 5%) of bentonite powder may reduce shrinkage and provide plasticity. Bentonite should be added dry to the cement-water slurry without first mixing it with water.

Some uncertainty exists as to whether bentonite should be used in the unsaturated zone, as bentonite may never fully hydrate and expand and it may subsequently desiccate and shrink. Improper hydration/dehydration could reduce the efficacy of the annular seal.[1,6,44] Bentonite may, however, rehydrate when wetted and so repair itself. Field tests of a bentonite slurry annular seal using a dye tracer in a semiarid environment indicated that bentonite retained its sealing characteristics in the vadose zone in an extreme temperature and low water regime.[13] It is uncertain if similar results would be obtained if the bentonite was emplaced in a pellet or chip form.

Bentonite, especially in the pellet form, has a proclivity to stick, hydrate, and bridge to any surface including protrusions from the well casing or the drill flights; for this reason, if pellets are chosen they should be emplaced by using a large, clean, dry funnel and a clean tremie pipe. The tremie should be rotated in a circular manner around the annulus as the tremie pipe is raised upward to ensure an even distribution of bentonite. If standing water is present in the annulus, then chips (or pellets) may have to be poured directly into the annulus from the surface.

All grouts (cement or bentonite slurry) must be uniformly mixed, and lumpy grouts should not be used. Grouts must be emplaced with a grout pump and a tremie pipe, which emplaces the grout from the bottom up. The tremie must discharge below the surface of the grout, with as little disturbance of the filter pack as possible. A weighted tape must be used to gauge the progress of grout emplacement, and the tremie should be raised at the same rate as grout emplacement.

Figure 3.12 A typical monitoring well.

Well completion details

A common monitoring well design incorporates a filter pack, fine sand above the filter pack, a bentonite seal from 2 to 5 ft thick above the fine sand, cement grout from the bentonite to the frost line (if applicable), and surface completion details. Surface completion details include a surface seal, a protective outer casing, a case-hardened steel lock, protective posts, and a well identification tag (Figure 3.12).

A surface seai consisting of either concrete or neat cement grout should extend from the top of the annular seal to ground surface (see Figure 12). In areas where frost heave could damage the well, the surface seal must extend a minimum of 1 ft below the frost line. The protective outer casing is set into the surface seal and extends below the frost line.

Figure 3.13 A monitoring well protected by a concrete pyramid.

The protective outer casing should be a minimum of 4 in. in diameter greater than the inner casing, and should be equipped with a lockable cover. The protective outer casing should have a 0.25-in.-diameter vent hole to preclude the build-up of potentially explosive gasses and to allow drainage of water. Sometimes, pea gravel is placed between the inner and outer casings to reduce the potential for frost damage and to reduce insect infestations. The inner well cap may need to have a hole drilled into it to allow the well to vent. Monitoring wells need to be protected from vehicle traffic, floods, vandalism, and even sabotage.[48]

A case hardened steel lock must be used to secure monitoring wells at all times. The lock serves to discourage vandalism and functions as a chain of custody seal. At no time should the well be left unlocked. Locks will most likely have to be replaced periodically, as lubrication of the locks with oil or graphite is not acceptable.

The well must be protected from damage from vehicles by using durable posts or a concrete pyramid (Figure 3.13). Special below-grade well vaults are available which allow traffic to drive on top of the well.

Each monitoring well should have an identification tag affixed to it with the well number, date of installation, well owner, driller, and top of inner casing elevation inscribed into it. This is accomplished by using a sharp awl or metal letter stamps to inscribe the required information onto a relatively thin (3 in. × 5 in.) aluminum plate . The aluminum plate is pop-riveted to the outermost casing or the inside top of the outer well cap.

Surveying the well

Depending on site objectives, two horizontal control points and at least one vertical control point (permanent benchmarks) at a site may be warranted. Benchmarks are established so the initial surveying is accurate, existing wells can be periodically resurveyed, and more wells can be readily surveyed into the network in the future, if they are needed.

Surveying is done in terms of a circuit, from a beginning control point to an end control point. The elevation can be brought in via an established benchmark such as bridges and section corners or in more remote areas with, at a minimum, three global positioning satellite (GPS) receivers. With GPS equipment, horizontal accuracy can be achieved from meters to submeters,[12] and vertical accuracy to a few hundredths of a meter may be possible. Site benchmarks need to be permanent and be located on the site map.

Monitoring wells need to be surveyed horizontally as well as vertically. The elevation of the north side of the top of the innermost casing of monitoring wells must be surveyed by a licensed surveyor to the nearest 1/100th of a foot above mean sea level, by reference to a National Geodetic Vertical Datum.[43] The north side of the innermost well casing should be surveyed, as the tops of inner well casings may be uneven. Having the surveyed elevation and subsequent water level measurements made on the north side of the innermost well casing as a convention will reduce errors resulting from water level measurements not being taken on the same side from event to event or on the same side that the well was surveyed.

Surveying consists of making multiple measurements, none of which are perfect. Errors involved with surveying include mechanical, human, and natural errors. Accuracy of horizontal measurements have been classified as being the first through fifth order; these are relative errors based on the length of the surveying loop. First-order accuracy is no more than 1/25,000, and fifth-order accuracy is no more than 1/1000. Obtaining first-order accuracy is considerably more expensive than fifth-order accuracy, and for most monitoring wells third-order accuracy (1/5000) may be sufficient.

Errors associated with surveying monitoring wells include improperly establishing the site benchmark(s) and/or improperly surveying individual wells. If the elevation of the site benchmark is in error, then each well in the network will be uniformly off in terms of other proximal benchmarks and, possibly, surface water features. Even if the site benchmark is off, the ground water flow direction and rate may still be *relatively* accurate in terms of the ground water flow direction within the site.

If individual wells are surveyed incorrectly, then ground water flow directions and rates will be in error, potentially causing serious misinterpretations of the hydraulics of the site, regardless of the accuracy of subsequent water level measurements in the wells. The author has seen surveying errors in excess of 30 vertical feet in individual monitoring wells. Establishing accurate elevations of monitoring points is very important, and even more so when the ground water gradient is relatively flat. Spot-checking the accuracy of the surveying work may be undertaken by using the well logs and a topographic map, or possibly resurveying suspect wells with another surveying crew.

Developing the well

After construction of the well a sufficient amount of time should elapse to allow the grout to set and cure. The amount of time will vary depending on the construction materials, methods, and climate; 72 hours should be considered the minimum amount of time allowed before subsequent activities are allowed to commence.

The next step for commissioning a monitoring well is to develop the well. The objective for monitoring well development is to remove foreign materials introduced during the drilling and installation of the well and to restore the natural hydraulic conductivity and water quality of the formation being sampled. Turbidity should be measured using a nephelometer during well development. In some geologic media, the USEPA limit of five nephelometric turbidity units (NTUs) or less may be difficult to attain (see Figure 4.1). If excessive turbidity is apparent in a monitoring well after development, then the documentation must be examined; if the well was erroneously designed or installed it may require additional redevelopment or replacement.

In fine-grained formation material such as silts and clays, bailing instead of surging or installing the well screen and filter pack in dry holes vs. wet holes yields samples with very low turbidity.[35] Wells constructed in fine-grained materials may have to be developed several times to reduce turbidity.

The use of prepacked screens in fine-grained media may also reduce turbidity problems, but redevelopment may be a problem. Pouring clean water down the casing at a relatively rapid rate, out the screen and into the filter pack, and up the open borehole prior to emplacement of the annular seal has also been suggested to reduce turbidity in formations with low hydraulic conductivity.[10]

The most accepted and commonly used methods for monitoring well development include surging with a surge block, bailing, pumping, or combinations of surging, bailing, and pumping. A surge block, similar to a piston, is attached to a rod which is raised and lowered in the monitoring well, thus forcing water back and forth through the screen. Some surge blocks have pressure relief holes and flexible "rings" (like automobile engine piston rings) are almost the same diameter as the well screen.

Numerous surge block designs are possible, and they can be purchased or fabricated by sandwiching inert, flexible gasket material between wooden disks (slightly less than 2 in. in diameter). A length of PVC casing filled with sand (with end caps) tied to a clean length of rope may adequately function as a surge block. Tubing attached to a pump or a check valve should be incorporated into the design of a surge block so that water will be brought to the surface simultaneously with surging.[32]

The surge block is raised and lowered, beginning at the top of the screen and then moving downward with increasing vigor, periodically removing fine material from the well with either a bailer or a sand-resistant pump. The force of surging depends on the speed and length of stroke. Surging can collapse the screen and casing and the surge block may become stuck ("sand locked") if proper techniques are not employed.

Bailing a well for well development is effective in relatively clean, permeable formations where the water flows freely into the borehole.[1] To develop a well by bailing, the bailer is agitated up and down at various depths of the well screen, and fine material is collected from the bottom sump with the bailer. Care must be exercised when allowing the bailer to free fall, when agitating, or during rapid removal so as to not exceed the collapse strength of the screen or casing.

The pump used for developing a well should be designed to hold up when pumping relatively large quantities of sand[28] and must be capable of operating at various pumping rates. A peristaltic pump may be preferred when developing shallow wells. Overpumping may require adequate disposal of contaminated development water.

When developing a well by overpumping, the pumping rate is substantially greater than the ability of the formation to deliver water. A variation of overpumping involves letting the pumped water column to fall back into the casing, causing a surging action. This technique is accomplished by starting and stopping a pump without a backflow device. A pump should not be allowed to operate in reverse or potentially contaminated air may be forced into the filter pack. The pump should be raised and lowered in the screened section during development to ensure that the entire screened interval is subjected to the overpumping and/or surging. Combining surging with overpumping is probably the best well development procedure.

Compressed air should never be used to develop a monitoring well due to potentially altering the ground water chemistry by introducing air (oxygen) and compressor oils into the screen, filter pack, and surrounding geologic media.[1]

Hydraulic conductivity testing methods

Subsequent to well development, a predetermined number of monitoring wells in the EMS should be tested to ascertain the hydraulic performance of the well and the hydraulic conductivity of the formation. If the well has not been properly constructed or developed, a pump test will not yield accurate data. Hydraulic conductivity is determined by using a slug test, a pump test, or, in highly permeable media, pressurization of the well.[38]

Although a pump test probably yields more accurate information about the hydraulic characteristics of an aquifer, pump tests on monitoring wells are often difficult to perform[10] so slug tests are more widely used. A field test should be used to determine hydraulic conductivity, as laboratory-derived hydraulic conductivities have been shown to vary by orders of magnitude from values derived by field methods.[10]

There are numerous slug tests and analytical methods used to assign a hydraulic conductivity to the proximal formation of a monitoring well. Figure 3.14[5] is a decision tree that may assist in choosing a slug test method applicable to the specific hydrogeologic conditions.[5]

A slug test is undertaken by quickly introducing or removing a solid object or a known volume of water into the well or by pressurizing the well. A change in head in the well of about a foot is generally adequate. A manual water level indicator, but preferably a continuous data logger, is used to record the drawdown or recovery of the water in the well (Figure 3.15). The response of the water level in the monitoring well is a function of the mass of water in the well and the transmissivity and storage coefficient of the aquifer.[5]

An analytical solution is applied to the time vs. drawdown (or recovery) data to determine hydraulic conductivity (or transmissivity and then hydraulic conductivity), which is used in turn to calculate ground water flow velocity. Numerous assumptions must be made specific to the slug test and analytical solution which may or may not be valid.

Documentation

Monitoring well documentation consists of the boring log, a complete and accurate well construction diagram including the survey information, the equipment decontamination report, a record of well development, the laboratory-derived vertical permeabilities, and an accurate, scaled site map showing the location of the well. If this information is not available, the "pedigree" of the well may be called into question.

Once a monitoring well has been constructed, developed, and slug tested, ground water sampling may commence. Depending on site objectives, base-line sampling consisting of three rounds of sampling (each round within approximately 14 days of the last) may be appropriate. Base-line sampling provides a basis for evaluating subsequent trends in ground water chemistry.

Drilling checklists

A standard checklist should be developed and used for ground water monitoring programs to ensure adequate workmanship and relative construction uniformity. Modifications to the following checklist may be appropriate. It should be the responsibility of the consultant and permittee to ensure compliance with other laws, ordinances, rules, and requirements when installing monitoring wells, including those specified in the governing well code. A checklist such as the following should be provided to the driller(s) when bids are requested.

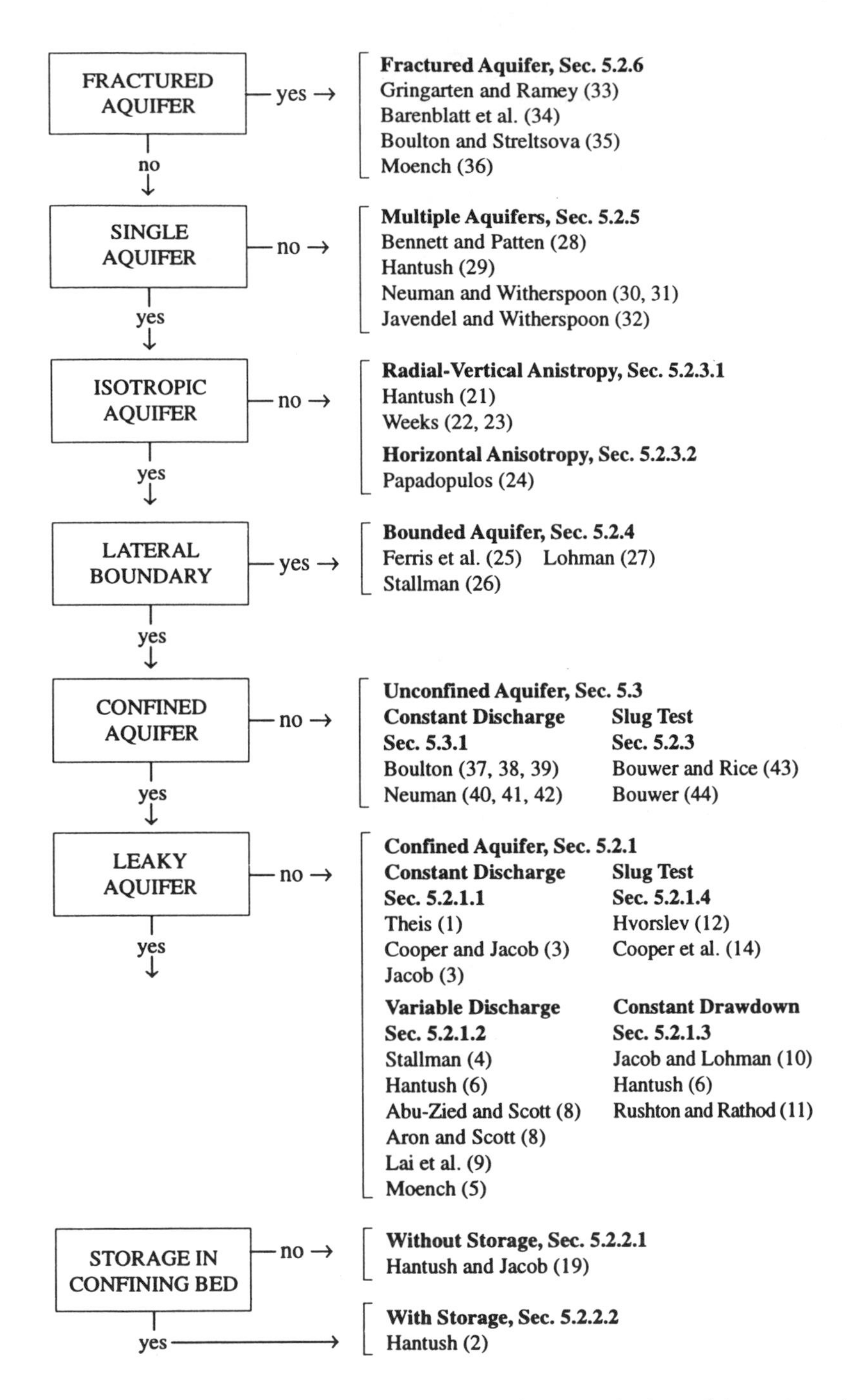

Figure 3.14 Analysis to determine appropriate slug test. (Adapted from American Society for Testing Materials (ASTM), D4043-91[5]).

Figure 3.15 Water level data logger. (Photograph courtesy of Geoguard, Inc.; *Medina, New York.*)

Monitoring well installation "bullet" checklist

1. The well driller must be a licensed driller in conformance with the appropriate governing agency.
2. In accordance with OSHA Regulation 1926.651, all utilities must be adequately notified where and when drilling is proposed.
3. Site-specific safety and health plan(s) in accordance with 29 CFR 1910.120 (b) (4) may be necessary, and this must be addressed in the work plan. A predrilling safety meeting should be held between the driller, consultant, and government regulators to define respective roles and responsibilities.
4. Wells proposed to be installed in flood-prone areas may require special design considerations, which should be addressed in the work plan.
5. The locations of monitoring points need to be considered in terms of regulatory program requirements, in addition to technical considerations.

6. The exact method of drilling needs to be specified in the work plan. Mud rotary drilling should be discouraged unless an alternative method is not possible. If air rotary or percussion drilling is selected, then foams must not be used and adequate filters must be used on the compressor so the potential for oil contamination is eliminated.
7. An individual with expertise in hydrogeology as defined by education and experience needs to be present on the drill rig to classify borings and to possibly facilitate field changes to the approved work plan. Changes in the field include more frequent soil borings, changes in methodology, etc. Field changes may need approval from regulatory agencies, and this should be addressed in the work plan.
8. When installing monitoring wells soil samples must be taken, at a minimum, every 5 ft or with each change in lithology using ASTM method D 1586 (split barrel), D 1587 (thin-walled tube), D 3550 (ring-lined barrel) or an approved equivalent method. ASTM D 2487 taxonomy must be used. Descriptions of the soil and lithologic units including textural classification, primary and secondary structures, voids, and soil color as per a Munsell chart must be included in the logs. In many drilling situations the most rapid change in lithology occurs in the first 15 ft so continuously logging the first 15 ft and the entire screened interval is recommended. In addition to anticipating when more frequent samples would be warranted in the work plan, the drill cuttings and the "pull" of the drill rig should be used to assess when more frequent soil samples are necessary.
9. Each soil unit identified should be laboratory tested for particle size, particle distribution, porosity, vertical permeability, and clay mineral content or cation exchange capacity.
10. The filter pack for the proposed monitoring well(s) needs to be of uniform, well-sorted grain size and be based on the cumulative percent retained (D-70) as determined from a grain size distribution curve.[1] The well intake (screen) size should only be determined after the filter pack is specified and should hold back between 85 to 100% of the filter pack material. In most situations this will require soil borings and laboratory analysis antecedent to monitoring well installation. The filter pack and screen size determinations need to be submitted to the regulatory agency prior to well installation.
11. The monitoring well screens and riser pipe must be either factory cleaned (certified as cleaned) and delivered to the site in plastic bags and removed from any potential source of contamination including gasoline, hydraulic fluids, etc., or adequately steam cleaned in the field. The work plan must specify which procedure is proposed to be used.

12. The filter pack and grout must be emplaced from the bottom up using a tremie and not poured into the annulus from the surface. If bentonite chips or pellets are not used for the annular seal, 2 ft of fine sand needs to be placed above the filter pack to prevent grout from invading the filter pack.
13. The type of annular sealant (i.e., bentonite or cement) and grout must be specified, including the type of bentonite proposed (i.e., chips, pellets, or a slurry) and the type of cement (i.e., Type I—V). Additives to the cement or grout must be approved by the appropriate regulatory agency.
14. Down-hole drilling equipment including auger flights or drill bits must be steam cleaned or high-pressure washed between drilling locations. The procedure must be specified in the work plan.
15. After the wells are installed, a minimum of 72 hours must elapse prior to well development to allow the grout to properly set up. A combination of surging with a surge block and/or overpumping is recommended for well development. To reduce the potential for altering ground water chemistry, compressed air must not be used for well development. The method of well development must be submitted to the appropriate regulatory agency for review and approval. Turbidity of 10 NTUs (nephelometric turbidity units) or less should be achieved in most monitoring situations.
16. Some of the newly installed monitoring wells may need to have the hydraulic conductivity ascertained (i.e., slug tested). The number and locations of well(s) to be tested, the test method proposed, and the analytic solution (including assumptions) must be submitted to the appropriate regulatory agency for review and approval prior to initiating the slug test.
17. Newer, more accurate permanent site bench marks need to be established if questions exist as to the accuracy and permanence of the existing site bench mark(s). This must be discussed in the work plan.
18. The north side of the innermost casing of the monitoring wells must be vertically surveyed to the nearest 1/100th of a foot, and the monitoring wells must be vertically surveyed (not using a measuring tape).
19. All monitoring wells in the monitoring system need to be permanently labeled, preferably with an aluminum plate approximately 3 in. × 5 in. pop-riveted to the outermost casing with the field number, date of installation, and well owner inscribed with an awl or metal stamps. Adequate well labeling must be addressed in the work plan.
20. The well log, construction diagram, formation log, laboratory analysis, surveying and slug testing results, and a (brief) written narrative discussing the drilling and installation of the monitoring well(s), including all deviations from the approved work plan, must be submitted to the appropriate regulatory agency for review and approval subsequent to well installation.

References

1. Aller, I., Bennett, T.W., Hackett, G., Petty, R.J., Lehr, J.D., Sedoris, H., Nielsen, D.M., and Denne, J.E., *Handbook of Suggested Practices for the Design and Installation of Ground-Water Monitoring Wells,* National Water Well Association, Dublin, OH, 1989.
2. ASTM D 4700-91, *Standard Guide for Soil Sampling in the Vadose Zone,* American Society for Testing and Materials, Philadelphia, PA, 1991.
3. ASTM D 1586-84, *Standard Method for Penetration Test and Split-Barrel Sampling of Soils,* American Society for Testing and Materials, Philadelphia, PA, 1984.
4. ASTM D 1587-83, *Standard Practice for Thin-Walled Tube Sampling of Soils,* American Society for Testing and Materials, Philadelphia, PA, 1983.
5. ASTM D 4043-91, *Standard Guide for Selection of Aquifer-Test Method in Determining of Hydraulic Properties by Well Techniques,* American Society for Testing and Materials, Philadelphia, PA, 1991.
6. ASTM D 5092-90, *Standard Practice for Design and Installation of Ground Water Monitoring Wells in Aquifers,* American Society for Testing and Materials, Philadelphia, PA, 1990.
7. ASTM D 5088-90, *Standard Practice for Decontamination of Field Equipment Used at Nonradioactive Waste Sites,* American Society for Testing and Materials, Philadelphia, PA, 1990.
8. ASTM D 2488-84, *Practice for Description and Identification of Soils (Visual-Manual Procedure),* American Society for Testing and Materials, Philadelphia, PA, 1984.
9. ASTM D 4220-89, *Standard Practice for Preserving and Transporting Soil Samples,* American Society for Testing and Materials, Philadelphia, PA, 1989.
10. Barcelona, M.J., Gibb, J.P., Helfrich, J.A., and Garske, E.E., Practical Guide for Ground Water Sampling, Illinois State Water Survey, SWS Contract 374, Champaign, IL, 1986.
11. Barcelona, M.J. and Helfrich, J.A., Well construction and purging effects on ground-water samples, *Environ. Sci. Technol.,* 20(11), 1986.
12. Burgess, R., Using GPS in a water well siting system, *Water Well J.,* October, 1993.
13. Butcher, L., Bentonite Slurry Grout in the Vadose Zone: An Effective Seal?, *National Drillers Buyers Guide,* April, 1993.
14. Cowgill, U.M., The chemical composition of leachate from a two-week dwell-time study of pvc well casing and a three-week dwell-time study of fiberglass reinforced epoxy well casing, *Ground-Water Contamination: Field Methods,* ASTM Spec. Tech. Publ. 963, American Society for Testing and Materials, Philadelphia, PA, 1988.
15. Driscoll, F.G., Ground Water and Wells, Johnson Filtration Systems, Inc., St. Paul, MN, 1986.
16. Edil, T.B., Chang, M.M.K., Lan, L.T., and Riewe, T.V., Sealing characteristics of selected grouts for water wells, *Ground Water,* May-June, 1992.
17. Gibs, J., Brown, G.A., Turner, K.S., MacLeod, C.L., Jelinski, J.C., and Koehnlein, S.A., Effects of small-scale vertical variations in well-screen inflow rates and concentrations of organic compounds on the collection of representative ground-water-quality samples, *Ground Water,* March-April, 1993.
18. Gibb, J.P., How Drilling Fluids and Grouting Materials Affect the Integrity of Ground Water Samples from Monitoring Wells, Forum, *GWMR,* Winter, 1987.

19. Gillham, R.W. and O'Hannesin, S.F., *Sorption of Aromatic Hydrocarbons by Materials Used in Construction of Ground-Water Sampling Wells,* ASTM STP 1053, American Society for Testing and Materials, Philadelphia, PA, 1987.
20. Godsey, K.D., The (sonic) wave of the future, *Water Well J.,* May, 1993.
21. Glotfelty, M.F., Monitoring well design considerations, *Water Well J.,* May, 1993.
22. Hewitt, A.D., Influence of Well Casing Composition on Trace Metals in Ground Water, U.S. Army Corps of Engineers, CRREL Spec. Rep. 89-9, April, 1989.
23. Hewitt, A.D., Leaching of Metal Pollutants from Four Well Casings Used for Ground-Water Monitoring, U.S. Army Corps of Engineers, CRREL, Spec. Rep. 89–32, September, 1989.
24. Hewitt, A.D., Potential of common well casing materials to influence aqueous metal concentrations, *GWMR,* Spring, 1992.
25. Hewitt, A.D., Dynamic study of common well screen materials, *GWMR,* Winter, 1994.
26. Hix, G.L., Squeaky clean drill rigs, *GWMR,* Summer, 1992.
27. Hix, G.L., Using mud rotary to drill monitoring and remediation wells, *Water Well J.,* May, 1993.
28. Hix, G.L., Monitoring well development: tools and techniques, *Water Well J.,* June, 1993.
29. Houghton, R.L. and Berger, M.E., *Effects of Well Casing Composition and Sampling Method on Apparent Quality of Ground Water,* Proceedings, The Fourth National Symposium and Exposition on Aquifer Restoration and Ground Water Monitoring, National Water Well Association, Dublin, OH, 1984.
30. Anon., ENV1091-10K, Johnson Environmental Well Products., St. Paul, MN, 1991.
31. Jones, J.N. and Miller, G.D., Adsorption of Selected Organic Contaminants onto Possible Well Casing Materials, *Ground Water Contamination,: Field Methods,* ASTM STP 963, American Society for Testing and Materials, Philadelphia, PA, 1988.
32. Kwader, T., Developing monitoring wells using surge blocks, a preferred method, *National Drillers Buyers Guide,* April, 1993.
33. Listi, R.L., Monitoring well grout: why I think bentonite is better, *Water Well J.,* May, 1993.
34. McLarty, F.R., Bentonite sealants and grouts, *National Drillers Buyers Guide,* May, 1993.
35. Paul, D.G., Palmer, C.D., and Cherkauer, D.S., The effect of construction, installation, and development on the turbidity of water in monitoring wells in fine-grained glacial till, *GWMR,* Winter, 1988.
36. Parker, L.V., Ranney, T.A., and Taylor, S., Softening of Rigid Polyvinyl Chloride by High Concentrations of Aqueous Solutions of Methylene Chloride, CRREL-SR- 92-12, U.S. Army Corp of Engineers, May, 1992.
37. Parker, L.V., Jenkins, T.F., and Black, P.B., Evaluation of Four Casing Materials for Monitoring Selected Trace Level Organics in ground Water, U.S. Army Corps of Engineers, CRREL Report 89-18, October, 1989.
38. Prosser, D.W., 1981. A method of performing response tests on highly permeable aquifers, *Ground Water,* 19(6), 1981.
39. Robbins, G.A., Influence of using purged and partially penetrating monitoring wells on contaminant detection, mapping and modeling, *Ground Water,* 27(2), 155, 1989.

40. Schalla, R., *A Comparison of the Effects of Rotary Wash and Air Rotary Drilling Techniques on Pumping Test Results,* Proceedings of the Sixth National Symposium and Exposition on Aquifer Restoration and Ground Water Monitoring, National Water Well Association, Dublin, OH, 1986.
41. Sikes, A.L., McAllister, R.A., and Homolya, J.B., Sorption of organics by monitoring well construction materials, *GWMR,* Fall, 1986.
42. Taylor, S., Parker, L., 1990. Surface Changes in Well Casing Pipe Exposed to High Concentrations of Organics in Aqueous Solution, U.S. Army Corps of Engineers, CRREL Spec. Rep. 90-7, March, 1990.
43. USEPA, RCRA Ground Water Monitoring Technical Enforcement Guidance Document, U.S. Environmental Protection Agency, Washington, D.C., 1986.
44. USEPA, Final Draft, Chapter Eleven of SW-846, Ground Water Monitoring, U.S. Environmental Protection Agency, Washington, D.C., 1991.
45. USEPA, Guide to Management of Investigation-Derived Wastes, Quick Reference Fact Sheet, Publ. 9345.3-03FS, U.S. Environmental Protection Agency, Washington, D.C., 1992.
46. Parker, L.V. and Ranney, T.A., Effect of concentration on sorption of dissolved organics by PVC, PTFE, and stainless steel wall casings, *GWMR,* Summer, 1994.
47. USEPA, Potential Sources of Error in Ground Water Sampling at Hazardous Waste Sites, Ground Water Issue, Report No. EPA/540/5-92/019, U.S. Environmental Protection Agency, Washington, D.C., 1992.
48. Hix, G.L., Monitoring well security, *Water Well J.,* November, 62–65, 1994.

chapter four

Static water levels and geochemical field parameters

I can foretell the way of celestial bodies, but can say nothing about the movement of a small drop of water.
Galileo Galilei, 1564–1642

Introduction

Monitoring wells are used to obtain information on ground water movement and ground water chemistry. By measuring the depth to water in monitoring wells in a network, the direction (and subsequently the rates) of ground water flow can be ascertained. Nonconservative characteristics of the ground water which must be measured in the field include temperature, specific conductance, pH, Eh (reduction-oxidation potential), dissolved oxygen, and alkalinity. These parameters assist in the physical and chemical interpretations of the site and are also useful for determining water stability prior to sample acquisition.

The following is intended to provide an overview on the equipment and techniques used for measuring static water levels and geochemical field parameters. Field parameters may be as important as the results obtained from the analytical laboratory, so the equipment selected should be of the best available quality and be used by individuals who are familiar with their operation.

Static water levels

The static water level in monitoring wells (and piezometers) is measured from the top of the inner well casing to the top of the undisturbed surface of water in the monitoring point. Obtaining static water levels in monitoring wells is necessary to determine ground water flow and may be used to determine the volume of water present in the casing for purging criteria.

To provide meaningful information on ground water levels the screen must be hydraulically sealed off from the rest of the formation, and the well

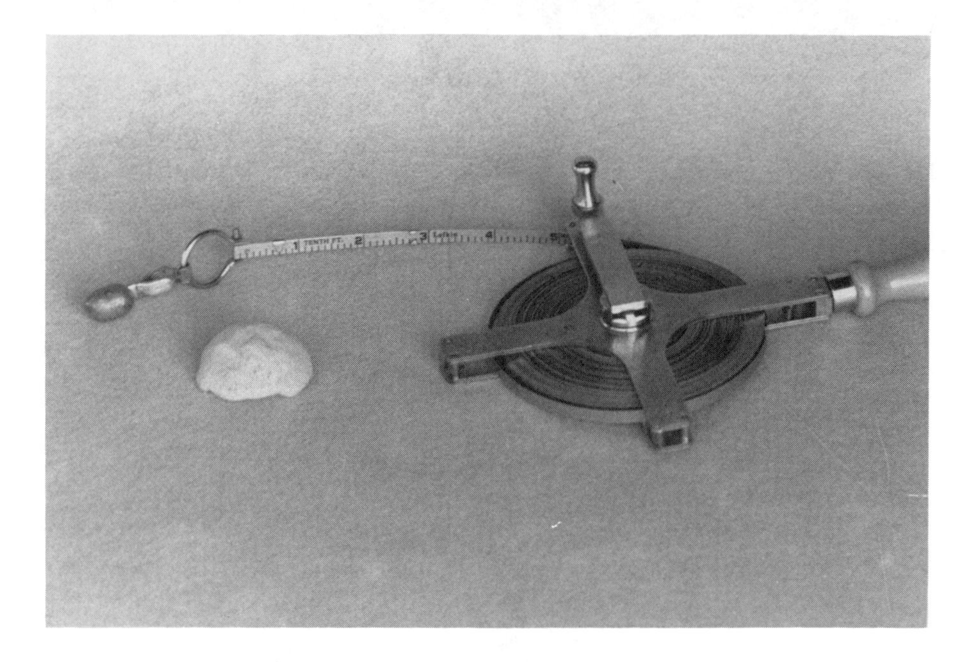

Figure 4.1 A steel tape and chalk for measuring static water levels.

should have been adequately developed. Water levels in uncased boreholes provide only a rough estimate of static water levels, and estimates are even more inaccurate in fine-grained material. Prior to measuring water levels any nearby wells should be turned off to reduce the potential for drawdown in the monitoring well.

Although other devices have been employed, the most widely used water level measuring devices are steel tapes or electronic water level indicators. A steel tape is commonly a 0.25-in.-wide black steel surveying tape on a hand-wound reel, and can be purchased in lengths of up to 500 ft. The black color of the back of the tape facilitates a visible contrast between the tape and chalk or indicator paste applied to the tape (Figure 4.1).

A slender, relatively heavy weight is attached to the end of it to keep the tape under tension; it has been recommended that the weight cause no more than a 0.05-ft rise of water in the well.[7] Lead fishing weights are commonly used as weights for steel tapes, but lead may be scraped off onto the casing. Stainless steel weights are preferable to lead weights if low levels of lead are of interest in the ground water.

Water level measuring devices should be checked before going into the field to ensure that they are intact and are in good working order. Water level measuring devices, even of the same make and model, may not provide the same water level measurement,[29] and periodic (at least once a year) checks on the accuracy or the length of the devices against a primary standard should be undertaken. Water level measuring tapes are read to the nearest one-hundredth of a foot and should not stretch more than 0.05% under normal use.[7]

As appropriate, an approved safety and health plan must be prepared and used for all laboratory and field activities (see Chapter 6). Toxic and explosive levels of gases and toxic or hazardous liquids may be present in monitoring wells. The use of suitable gas monitoring equipment and appropriate clothing including disposable gloves, safety glasses, etc., should be addressed in the site-specific safety and health plan, which would encompass taking static water levels.

It should be a convention that the north side of the innermost casing be used to take liquid levels in all monitoring wells to avoid inconsistencies from sampling event to sampling event due to uneven casing ends. Knowing from previous experience the anticipated water level in the well can reduce the time required to obtain a static water level when using a steel tape; otherwise, several attempts may be required.

Most steel tapes are only calibrated in hundredths of a foot at the end of the tape, so it may be necessary to carry along a small piece of steel tape which has the 0.01-ft graduations. To obtain a static water level using a steel tape, chalk or indicator paste is applied to the bottom 2 ft of the tape and the tape is carefully lowered to the anticipated static water depth. The tape should be stopped at a predetermined stop mark on the tape close to the anticipated water level (for example, stop lowering the tape at the 35-ft mark on the tape). The stop mark on the tape should be recorded and the tape brought back to the surface. It is important to only lower the tape to the stop mark until ready for retrieval (never lower past the stop mark and pull up to it) or else a false reading will be given to the water level.

The difference between the predetermined stop mark on the tape (in this example 35 ft) and the part of the tape where water has washed off the chalk or changed the color of the indicator paste is the depth to ground water from the measuring point of the well, i.e., the north side of the innermost casing. The north side of the innermost casing should have been previously surveyed, so knowing the depth to ground water distance can be quickly converted to a static water level in terms of absolute elevation (i.e., NGVD or MSL). The depth to the bottom of the well should also be determined to evaluate potential silting in of the well, to determine the volume of water in the casing, and to ensure that the correct (nested) well is being sampled.

Using a steel tape in subfreezing weather (the water freezes to the tape) or in the rain (the indicator is washed away) can be a challenging task. Steel tapes sometimes break, and instead of discarding the tape, it can still be used if a separate length of steel tape marked in 0.01-ft calibrations is used alongside the broken tape and care is exercised to ensure that the broken portion is accounted for in water level measurements. The chemistry of the chalk or indicator paste may also be of concern in some monitoring situations.

Electric water level indicators may be as accurate as a steel tape, especially those that use a steel tape enclosed in Teflon™. Electric water level indicators use ions in the water to close a circuit which activates a tone and/or an indicator light, and the depth to the water surface is read from the tape (Figure 4.2). To get an accurate reading with an electric tape, raising and

Figure 4.2 An electronic water level indicator. (Photograph courtesy of Keck, Inc.; *Williamston, Michigan.*)

lowering the tape a few inches a couple of times may be necessary after the probe enters the water. Some electronic water level indicators have a tip which protrudes beyond the actual sensor, and this length must be accounted for when taking the depth to the bottom of the well. It is a good practice to take two measurements of the static water level for each well, and measurements should agree within 0.02 ft.

If a light nonaqueous phase liquid (LNAPL) such as gasoline is present in the well, an electric circuit cannot be completed and the indicator tone and/or light will not function. Special bailers, steel tapes with indicator paste, electronic interface indicators, or a float device that attaches to the electric water level indicator allows a determination of the depth and thickness of a LNAPL layer. Intrinsically safe (and usually grounded) equipment should be used in the presence of potentially explosive levels of LNAPL (e.g., a gasoline phase).

An interface probe uses an optical sensor to determine if the sensor is in a liquid, and a conductivity meter to determine if it is in water (Figure 4.3). The LNAPL/air interface reading should be taken first to prevent dripping LNAPL from enhancing the reading. The LNAPL/water reading is best taken by quickly lowering the probe through the LNAPL layer and taking the reading from the water to the LNAPL interface to prevent coating the conductivity probe.[35]

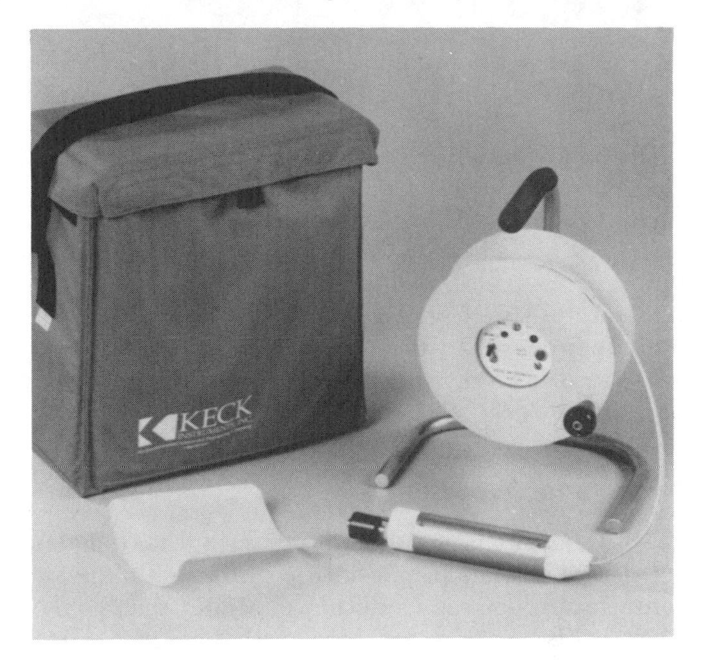

Figure 4.3 An electronic interface tape. (Photograph courtesy of Keck, Inc.; *Williamston, Michigan.*)

There is another method to determine LNAPL thickness once the depth to liquid is determined: a special bailer equipped with a flapper valve is slowly lowered through the product and brought to the surface and the thickness is measured in the bailer. This method may not be as accurate as the other methods due to the operation of the bailer. An interface meter, a coated tape, or bailer may also be used to determine the presence and thickness of a dense nonaqueous phase liquid (DNAPL) in the bottom of the well.

A poor correlation between the oil layer thickness in sediment cores and in monitoring wells has been observed,[16] and grain size distributions should be used to assist in LNAPL volume determinations.[26] Correction factors should be applied to ground water elevations where LNAPL is present, as the presence of LNAPL can distort (depress) static water level elevations.[25] This correction is given by:[46]

$$CDTW = DTW - (PT \times G)$$

where

CDTW	=	corrected depth to water
DTW	=	measured depth to water
PT	=	product thickness
G	=	specific gravity, given by G = (density of free product/ density of water)

Taking water levels in domestic or municipal wells can be a difficult task, and "hanging up the tape" in the well can be a problem. In extreme cases the pump may have to be pulled, so taking water levels in domestic or municipal wells should be done carefully, with as much information as possible known ahead of time about the well and pump.

Between monitoring wells, and especially between domestic wells, water level indicators must be adequately cleaned to reduce the potential for cross contamination and the spread of bacteria that can plug a well and impart a bad taste to the water. This may be accomplished by washing the wetted end of the tape in a nonphosphate detergent followed by an isopropyl alcohol rinse (see Chapter 6). The cleaning method needs to be documented in the field notes.

Geochemical field parameters

After taking static water levels in monitoring wells, the well may be purged until stabilized, while simultaneously taking geochemical field parameters including temperature, specific conductance, pH, Eh, dissolved oxygen, and turbidity. There has been a great deal of research done on if and how purging should be undertaken; Chapter 5 examines purging criteria and the types of equipment used to collect ground water samples (e.g., pumps and bailers).

Measuring geochemical field parameters is necessary for an accurate hydrogeochemical interpretation of the site, and the parameters may be used for well stabilization criteria. Equipment and techniques used to obtain stabilization parameters should be USEPA-approved equivalent methods.

Obtaining geochemical field parameters has been accomplished by taking an aliquot of the ground water and inserting the probes into the subsample, placing the probes in the bucket into which the water is being pumped, or using a flow-through cell (Figure 4.4). A multiport flow-through cell is required to obtain reliable geochemical field parameters; it allows the probes to be in contact with the sample water but keeps the sample water from coming into contact with the atmosphere, which could rapidly alter the sample. A flow-through cell must never be pressurized or damage to field parameter probes may result.

A multiport flow-through cell may be fabricated or purchased. A clear Plexiglas®, conical interior design with a volume of 700 mL has been suggested to allow viewing of the sample water and the probes and to prevent a "streaming potential" by the pH electrode.[20] A "streaming potential" is a static charge induced by pure water moving through the tubing and the cell.[21] Any gas bubbles that adhere to the probes can be removed by gentle tapping. Multiport flow-through cells may need to be immersed in a bath of the purge water to reduce ambient temperature influences on the sample.

Temperature

The temperature of the ground water is necessary for determining saturation and stability with respect to ions and products of reactions and may also

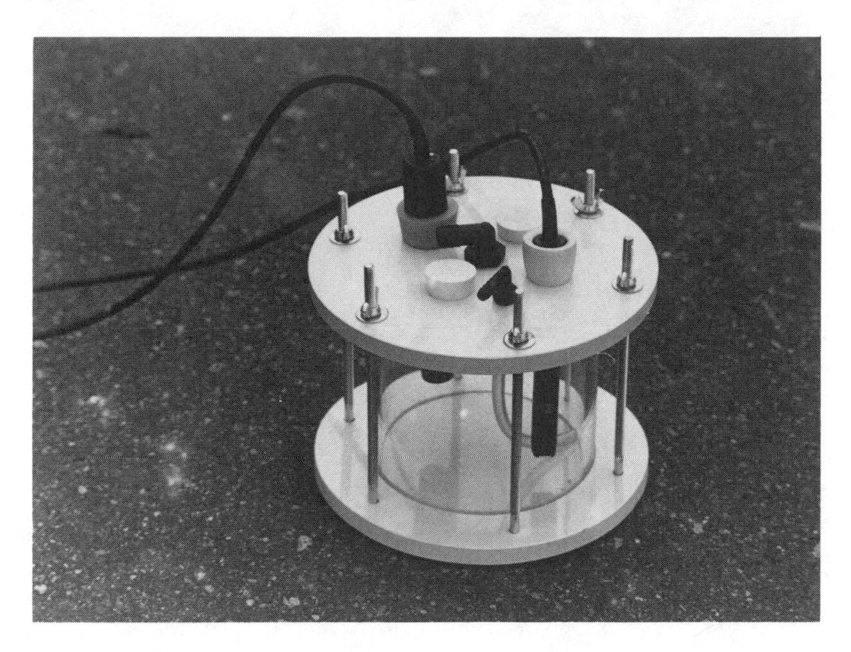

Figure 4.4 A fabricated multiport flow-through cell.

provide insight on the residence time of ground water in an aquifer. Mercury-filled glass thermometers or digital thermometers may be used for field information. The thermometer should provide resolution to at least 0.1°C[1] and should be routinely checked against a thermometer certified by the National Institute of Standards and Technology.[43] Temperature is not a good primary indicator for determining well stability during purging.

pH

The two main classes of chemical processes in ground water are oxidation-reduction reactions that involve the transfer of electrons and acid-base reactions that involve the transfer of protons. The pH of a solution is defined as the negative logarithm of the hydrogen ion activity (activity is approximately the concentration) in moles per liter, or the intensity of the acidic or basic character of a solution at a given temperature. Major constituents in ground water are in the range of 10^{-4} mol/L and up. The hydrogen ion concentration is very low for solutions that are not strongly acidic, and at pH 7 only 1×10^{-7} mol/L of hydrogen ions are present. The neutral point in a solution is temperature dependent; for example, at 25°C a pH of 7 is neutral, and at 0°C a pH of 7.5 is neutral.

One of the most important reactions controlling pH in a natural water system is the reaction of carbon dioxide with water. Some reactions, notably those involving ferrous and ferric iron, form precipitates that influence pH, and most reduction-oxidation (redox) reactions are pH related. The pH and redox potential greatly influence geochemical equilibrium and solubility of many species including metals in ground water. Many metals in ground

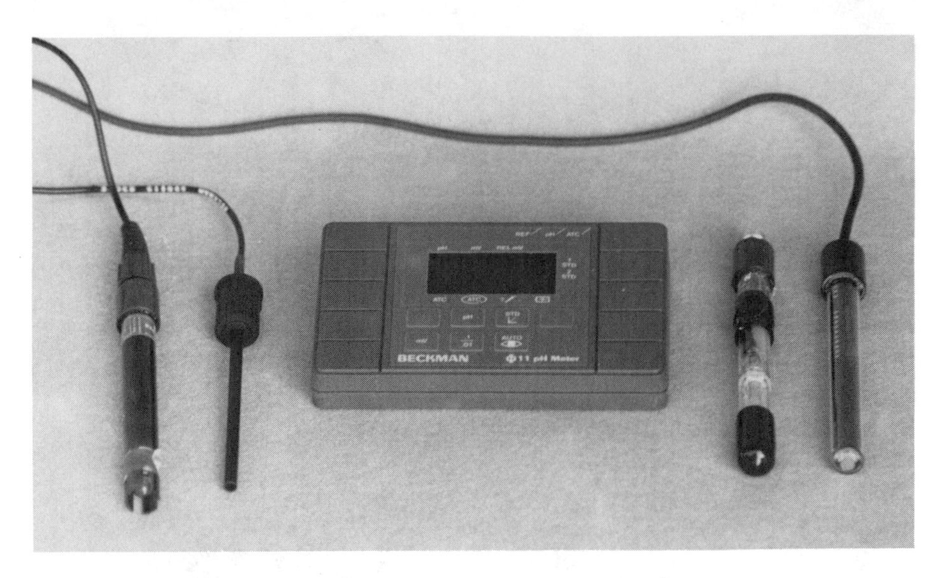

Figure 4.5 Left to right: pH probe, temperature probe, pH meter, and Eh probes. (Photograph courtesy of Beckman Scientific, Inc.; *Fullerton, California.*)

water exhibit amphoteric solubility, where metals may be mobilized at both low and high pHs.

Comparing horizontal and vertical changes in pH assists in characterizing the ground water and determining contaminant migration and chemical speciation. Changes in pH have been found to be relatively insensitive to purging[12] and so should not be used as a primary indicator of well stability prior to sample acquisition.

Electrodes that measure pH rely on a transfer of ions (ion exchange) to produce a current between the solution to be tested and the electrode, which can then be displayed on a meter. A standard glass membrane electrode consists of a Na-Ca silicate bulb filled with a HCl solution of known hydrogen ion concentration. A reference electrode, consisting of a silver wire coated with silver chloride, is immersed in the HCl solution. The glass acts as a semipermeable membrane, with sodium ions on the outer surface of the glass bulb being readily leached and replaced by hydrogen ions, which creates an electrical imbalance. The reference electrode resolves the charge imbalance in the HCl solution by generating electrons, which are then measured as voltage (Figure 4.5).[32]

A pH meter is generally accurate to ±0.1 pH unit. The pH meter must be manually or automatically compensated to correct for the effect of temperature on the instrument. This compensation does not, however, adjust the pH to a common temperature, so the temperature of the sample should be reported for each pH measurement.[5] Analog meters (pH, specific conductance, dissolved oxygen) may need to be calibrated in the same position (e.g., lying flat or set upright) as when used, but this is not required for meters with digital displays.

Some pH probes require immersion in water for several days prior to use in order to form a "gel layer". A pH meter is subject to drift and requires, at minimum, a daily two-point calibration utilizing two calibrating solutions that bracket the range of pHs anticipated in the field. Buffer solutions should have a shelf life of about 3 months, and the buffer solutions should be warmed or cooled to within 2°C of the water to be tested. To calibrate the meter, place the probe in the first buffer solution and gently stir. Empty the container and place new buffer solution into the container and read the pH. Repeat this process until two successive readings are within 0.02 pH units. Wash the electrode with deionized water, and repeat the above process with the second buffer solution.

When using a pH meter without a flow-through cell, the sample requires gentle stirring to reduce carbon dioxide buildup and ensure a stable probe reading. In a dilute, poorly buffered solution it may be necessary to equilibrate the probe by immersing it in three or four successive portions of the sample and then testing a fresh aliquot of the sample. At very high or low pHs the accuracy of a pH meter is decreased.

The rate of flow through the flow-through cell must be low enough to prevent inducing a "streaming potential" voltage. The manufacturer's instructions need to be followed. A pH probe is usually stored in either tap water or a KCl solution, and probes generally last between 12 to 18 months before they require replacement. Affixing water-resistant tape to the top of the probe with the date of initial use written on it in indelible ink is recommended. The tip of the probe should not be handled, and pH probes should not be exposed to temperatures below freezing. If a hole in the probe is used to fill it with KCl, then the hole should be left open during use, and nothing other than the filling solution should be allowed into the hole.

When calibrating a pH meter or taking a pH reading outside of a flow-through cell, the solution should be gently stirred; stirring should be discontinued just before reading the pH. Follow the manufacturer's instructions to clean a probe with a soap-and-water wash, followed by several rinses of water, followed by immersing the lower part of the probe in 10% HCl to remove any film on the probe. The probe should then be rinsed several times with deionized water before being used.

Errors of 0.1 to 0.2 units in pH measurements may introduce errors of a factor or two to an order of magnitude when the saturation states of solids are calculated. Differences as large as 2.8 pH units have been observed between measurements taken in the field and samples measured in a laboratory after 5 to 120 days, and large instantaneous errors in pH were shown to result from measuring pH in open conventional systems with waters not in equilibrium with atmospheric carbon dioxide.[36]

In addition to pH intensity, pH buffering of the system must be considered. A solution is said to be buffered if the pH is not greatly altered by the addition of moderate quantities of an acid or base. Alkalinity is a measure of pH buffering. Alkalinity needs to be determined in the field, and it will be addressed in greater detail later in this chapter.

Specific electrical conductance

Electrical conductance or conductivity in water has been defined as "the reciprocal of the resistance in ohms (mhos) measured between opposite faces of a centimeter cube of an aqueous solution at a specified temperature."[4] Alternating current must be used for determining specific electrical conductance to prevent a polarization of the electrode surfaces.[1] Conductivity in water depends on the concentration of ions, mobility, valence, relative concentrations, and the temperature of the solution.

Specific conductance corrects the conductivity measurement to 25°C; in dilute solutions of most ions, an increase of 1°C increases conductivity by about 2%.[23] Natural waters have specific conductances that are usually much less than 1 mho; to avoid using decimals, data are usually reported as micromhos per centimeter, or just micromhos (μmhos). The International System of Units measures specific conductance in milliSiemens per meter (mS/m).

The presence of charged ionic species allows current to be conducted through the solution, but other undissociated species in the solution complicate the relationship between specific conductance and total dissolved solids. An estimate of total dissolved solids in mg/L can be made by multiplying specific conductance in μmhos by a factor of between 0.55 to 0.9, depending on the concentration of soluble component.[1] Specific conductance, as a field parameter, may best reflect major ion chemistry in ground water as it is less affected by either volatilization or small-scale heterogeneity.[12]

In addition to providing an estimation of total dissolved solids, comparing upgradient- and downgradient-specific conductances can provide a valuable first approximation of the mobility of the ionic species in the system. Specific conductance appears to be a good index for well stabilization prior to sampling for certain chemical constituents (notably inorganic species) in ground water.

A typical conductivity meter has a platinum electrode that requires replatinizing when the readings become erratic, when a multipoint calibration shows discrepancies between actual and measured concentrations, or when an inspection of the probe shows that any platinum black has flaked off. The manufacturer's recommendations for cleaning the instrument should be followed, and a cleaning solution consisting of 1 part by volume isopropyl alcohol, 1 part ethyl ether, and 1 part HCl (1 + 1) may be suitable for cleaning the probe.[4]

A conductivity meter should be checked over the range of conductivities anticipated in the field. Potassium chloride solutions of various conductivities can be obtained commercially or can be prepared by adding 745.6 mg of 0.01 *N* anhydrous KCl to 1000 mg/L of distilled deionized water at 25°C to make a solution with a conductivity of 1413 μmhos/cm. By carefully diluting the solution, various conductivities can be obtained. Precision of commercial conductivity meters is typically between 0.1 to 1.0%, and a reproducibility of 1 to 2% after calibration should be expected.[1]

Reduction-oxidation potential

Reduction is defined as the gain of electrons, and oxidation as the loss of electrons. Reduction-oxidation (redox) reactions involve the transfer of electrons between reactant and product assemblages. Free electrons do not exist in solution, so an oxidation reaction (loss of electrons) must be balanced by a reduction reaction (a gain of electrons). Redox potential (Eh) is defined by the Nernst equation and is the energy gained in the transfer of 1 mol of electrons from an oxidant to H_2.

The activity (concentration) of electrons, analogous to pressure, does not correspond to a concentration because there is no net production or consumption of electrons in a redox process.[14] Due to the transfer of electrons, a change occurs in the oxidation states of the reactants and products (solid, dissolved, and gaseous). The dominant aquifer redox reactions involve aqueous and solid species, mediated by bacteria.

The most important species that undergo redox reactions in ground water involve carbon, iron, sulfur, and nitrogen. Ground water is usually isolated from atmospheric oxygen so it tends to be depleted of free oxygen (i.e., reduced). Reduced organic matter is usually expected to be the electron donor, but reduced inorganic species may also contribute electrons.[22]

The introduction of organic loading from landfill leachate, nonaqueous phase liquids, etc., typically leads to the development of reduced ground water zones, as the organic material is microbially degraded. The development of reduced zones depends on the availability of oxidized species in the aquifer matrix; methods for determining the oxidation capacity of aquifer sediments have been reported.[11,22]

Bacteria catalyze nearly all of the important redox reactions in ground water. Although reactions may be thermodynamically spontaneous, they may require a microbial catalyst for them to proceed at an appreciable rate. Aerobic, anaerobic, or facultative bacteria in ground water systems, ranging in size from 0.5 to 3 μm, utilize electrons for their metabolic processes. In the absence of catalyzing bacteria, some redox reactions could take hundreds of years to reach equilibrium.

Free oxygen, nitrate, sulfate, and ferric iron are important electron acceptors in aquifers. For reduction of inorganic constituents to occur, organic matter also is generally oxidized in ground water systems. As the oxidizing agents (constituents that undergo reduction, i.e., gain electrons, such as O_2, NO_3, $Fe(OH)_3$ and SO_4) are consumed, the ground water becomes more and more reduced. Limiting factors in a sequence of redox reactions from aerobic to anaerobic include:

- Having organic matter in a consumable form throughout the process
- Having the necessary macro- or micronutrients for the mediating bacteria
- Having relatively stable temperatures in the ground water so the biochemical processes are not disrupted

If any one or more of these factors is limiting, some redox reactions may be prevented from proceeding from beginning to end.[18]

Redox buffering (analogous to pH buffering, but referring to O_2, NO_3, SO_4, etc.) may enable the redox potential to remain stable until the free oxygen is depleted and then cause it to drop abruptly. Following the consumption of free oxygen, the redox potential remains relatively constant until other species such as nitrate or sulfate are consumed, and then another drop may occur. If the redox potential is positive, it has been suggested that dissolved oxygen concentrations be measured; if negative, then hydrogen sulfide and ammonia should be measured.[17]

Redox measurements on natural waters are difficult to make and interpret due to chemical and/or electrochemical disequilibrium.[24] Measured redox potentials are usually 400 to 600 mV lower than calculated redox potentials because dissolved oxygen does not exert its full potential on the redox electrode.[19] Electrode poisoning (from organic matter, sulfide, and bromide), lack of internal equilibrium, presence of multiple redox couples and inert redox couples, and irreversible reactions cause the calculated and measured Eh values to rarely correspond.[1] Oxygen reduction may involve hydrogen peroxide as an intermediate in oxic environments, so a comparison of calculated to measured Eh may not be appropriate.[10]

Redox potentials in natural waters are usually mixed potentials, which are impossible to relate to a single dominant redox couple in solution.[27] Various redox couples in natural waters are not in equilibrium with each other, so a single redox measurement to characterize oxygenated waters may not have any meaning.[38] Due to the nonuniform distribution of bacteria in an aquifer matrix, it is possible to have different redox potentials in relatively close proximity to each other.

Instead of using the redox potential as a "master variable" in geochemical interpretations, considering it in terms of a redox sequence or redox zones along the ground water flow path may help identify the major redox couples that control the redox potential. This sequence can be described as follows:[13,15]

1. Oxic (oxygen rich): redox potentials of +250 mV to +100 mV
2. Post-oxic (iron rich): redox potentials of +100 mV to 0 mV
3. Sulfidic (sulfate rich): redox potentials of 0 mV to –200 mV
4. Methanic (methane rich): redox potentials below –200 mV

The rates of redox reactions and the transport of the products of reactants will determine whether or not idealized redox zones will be observed. Vertical redox gradients of more than –40 mV/m depth, which were at least an order of magnitude greater than those reported along the general path of ground water flow, have been observed.[10]

The relationship between Eh and pH determines the stability and occurrence of many ions and compounds. For example, in a dissolved iron system with dissolved iron at 10^{-6} mol at standard conditions and a pH of 6 and an Eh of –200 mV, Fe^{2+} (ferrous iron) is stable and quite soluble; however, with a pH of 6 and an Eh of +400 mV, $Fe(OH)_3$ (ferric hydroxide) is stable. Eh-pH stability field diagrams provide stable products of redox reactions, and if different elements such as sulfur are present, different (stable) compounds are possible.

With a pH of between 5 and 9 and an Eh of +200 mV to –100 mV, a considerable amount of ferrous (very soluble) iron can be present in ground water. Precipitation of ferric hydroxide can lead to large (50 to 100%) decreases in the dissolved concentrations of some trace anions and cations including lead, cadmium, zinc, arsenic, vanadium, phosphate, and probably other trace metals that are adsorbed to freshly formed ferric hydroxide particles.[37] Arsenic, cadmium, chromium, copper, iron, mercury, molybdenum, nickel, lead, selenium, uranium, vanadium, and organic compounds such as polychlorinated biphenols, trichloroethane, benzene, and toluene all have multiple redox (valence) states, and the same compound may be more (or less) toxic depending on the redox state.[28]

At redox equilibrium the potential difference between the ideal indicator electrode and the reference electrode is the redox potential of the system.[2] Indicator electrodes are usually made of platinum; the reference electrode is commonly made of silver/silver chloride.

Measuring redox potential in the field may be undertaken using redox probes with a pH meter (see Figure 4.5). Redox potential is measured in millivolts. The performance of Eh probes is measured by placing the indicator and reference electrodes in Light's solution or Zobell's solution. Light's solution is 48.22 g of ferrous ammonium sulfate dissolved in 56.2 mL of sulfuric acid in water, diluted to 1 L. At 25°C with a silver/silver chloride reference electrode, Light's solution has a potential of +438 mV.[3]

Zobell's solution (which may not be as stable or potentially useful as Light's solution) is 3×10^{-3} *M* potassium ferrocyanide and 2×10^{-2} *M* potassium ferricyanide in 0.1 M KCl. At 25°C, Zobell's solution has a potential of 428 mV. Redox measurement is made by comparing the reference solution Eh and sample Eh at the same temperature.[1] Eh of the sample (using Zobell's solution) is given as:

$$Eh(sam) = Eh(obs) + Eh(Zref) - Eh(Zobs)$$

where:

Eh(sam) is the redox potential of the sample relative to the standard hydrogen electrode
Eh(obs) is the measured (observed) potential of the sample
Eh(Zref) is the theoretical Eh of the reference electrode and reference solution
Eh(Zobs) is the measured (observed) potential of the reference solution[1]

If drift or erratic behavior is observed when placing the electrodes in a standard solution, the electrodes should be cleaned, refilled, regenerated, or possibly replaced. If the instrument is off by ±10 mV when using the standard solution, follow the manufacturer's instructions to polish the indicator electrode with 400- to 600-grit wet/dry carborundum paper, and check again. If ±10 mV cannot be attained, then another probe should be used. If the Eh increases sharply when a small amount of dilute NaOH is added to a sample, then the polarity is reversed and this condition must be corrected.

When measuring, the solution should be gently stirred, and it may take several minutes for the probes to equilibrate with the sample. Successive readings of ±10 mV in the same solution over 10 min is considered adequate. If a KCl-saturated calomel electrode is used as a reference electrode, 242 mV at 25°C must be added to convert to true Eh.[28]

Platinum electrodes may have to be carefully pretreated according to the manufacturer's instructions and stored in an oxygen-scavenging solution of 0.2 *M* sodium sulfite, as the probes may be sensitive to the presence of oxygen. Routinely changing redox probes after exposure to water containing oxygen should be undertaken, as variations of several hundred mV have been observed between fresh and used electrodes, depending on the time since their last exposure to oxygen.[46] The immersible ends of the probes should be stored in water between measurements, and the fill-holes of reference electrodes should be covered when not in use to prevent evaporation.[3]

Awareness of the relationships between pH and Eh and the concepts of pH and redox intensity and buffering is necessary to assess contaminant mobility and attenuation. Identification of the major redox couples present in the ground water combined with accurate redox measurements may permit a generalized understanding of the chemical processes and a systems approach toward equilibrium.

Dissolved oxygen

Dissolved oxygen concentrations have a significant effect on the potential for degradation of organic contaminants and the mobility of trace metals in the ground water. The solubility of oxygen increases with increasing hydrostatic pressure and decreasing temperature. Dissolved oxygen concentrations greater than 1.0 mg/L have been measured in samples derived from relatively deep ground water.

In biochemical reactions, dissolved oxygen is always consumed and is never produced. Dissolved oxygen concentrations of 0.05 mg/L are considered limiting for most aerobic bacteria,[18] and the types and numbers of bacteria present (aerobic, anaerobic, or facultative) control the potential for degradation of organic compounds. Most alkyl benzene and chlorobenzene groups are probably stable in anaerobic water but may be biodegradable in aerobic water. Trichlorethylene is stable in aerobic water and potentially biodegradable in anaerobic water. Of 16 inorganic parameters with maximum contaminant levels (MCLs), 9 have multiple oxidation states and therefore are sensitive to changes in dissolved oxygen concentrations.[34]

A three-dimensional dissolved oxygen profile around contaminant zones and the surrounding region should be considered for the investigation of contaminant migration. Dissolved oxygen concentrations may provide insight

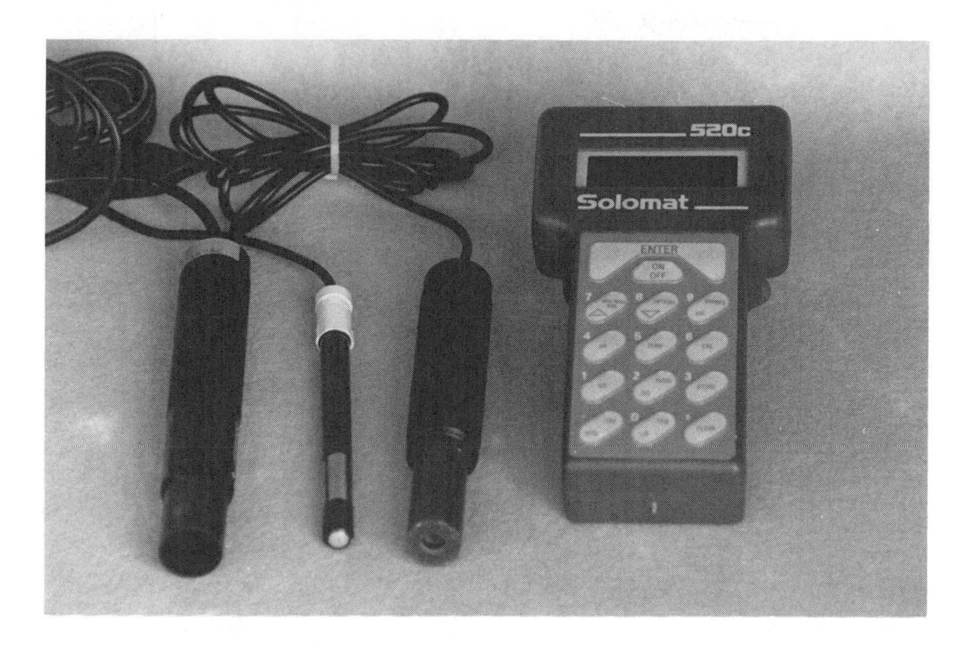

Figure 4.6 Instrumentation for measuring D.O., turbidity, and specific conductance. (Photograph courtesy of Solomat, Inc.; *Norwalk, Connecticut.*)

on the residence time of ground water, the potential for contaminant migration and attenuation, and the stability of water being purged from a well.

Dissolved oxygen concentrations can be determined in the field by using a probe or a titration method. Accurate dissolved oxygen concentrations cannot be derived from redox potentials. A probe allows dissolved oxygen to diffuse across a semipermeable membrane into an electrolyte solution. A current is applied to the electrolyte causing it to be reduced, and the current produced is proportional to the dissolved oxygen concentration (Figure 4.6).

The rate of oxygen diffusion across the semipermeable membrane is influenced by temperature, barometric pressure, fouling of the membrane, salinity of the sample, and whether adequate replenishment of sample water in proximity to the probe occurs. Due to changes in molecular activity, approximately a 3% increase or decrease in diffusion of oxygen through the membrane will occur for each degree Celcius change, so a dissolved oxygen meter should be equipped with automatic temperature compensation circuitry. Changes in elevational or barometric pressure will also influence mechanical dissolved oxygen measurements, and a dissolved oxygen meter may have to be corrected several times a day in the field for changes in elevational or atmospheric pressure.

Oily fluids may foul the semipermeable membrane, and gases including hydrogen sulfide, halogens, and nitrous and nitric oxide may cause interferences with measurements. Dissolved oxygen meters may be corrected for salinity either with a chart (linear interpolation) or by adjusting the meter itself. An adequate supply of sample water must be furnished to the probe

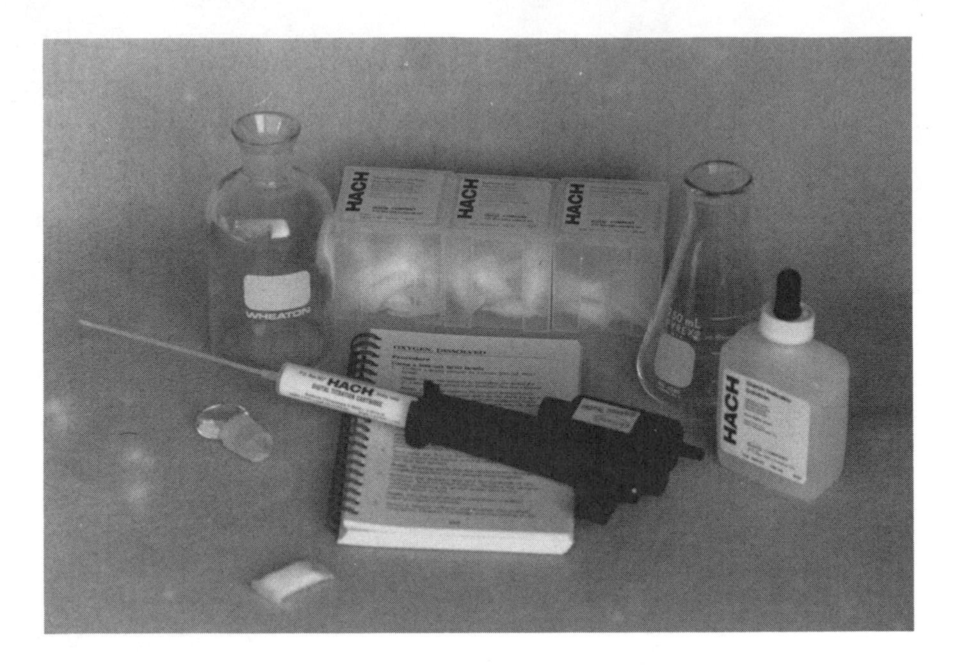

Figure 4.7 Titration equipment for determining dissolved oxygen in water. (Photograph courtesy of Hach, Inc.; *Loveland, Colorado.*)

as the water in the immediate area of the probe becomes depleted of oxygen as it diffuses across the membrane. Some dissolved oxygen meters are equipped with a stirring unit, which replenishes water to the probe.

Titration using the Winkler Method[41] or an equivalent method should be used in conjunction with a dissolved oxygen meter at least once a day to determine the probe's performance. Titrating is a labor intensive process but may be more accurate, depending on the proclivity of a probe to drift.

Most methods for determining dissolved oxygen in the field either entrain oxygen as part of the sampling process (especially at very low levels of dissolved oxygen) or the method itself may have a practical quantitation limit of 0.1 mg/L. Precision and detection limits in the field for both the titration and electrode methods are approximately 0.2 mg/L.[34]

Titrations do not provide a continuous reading; however, newer, easier titration methods using premeasured quantities of reagents may provide excellent dissolved oxygen concentration resolution in the very low range (1 to 10 μg/L), but may need USEPA approval as an equivalent method (Figure 4.7).

Turbidity

Turbidity is an expression of the optical properties of the sample which causes light rays to be scattered and absorbed rather than being transmitted in straight lines through the sample.[7] Turbidity is caused by dissolved or suspended particles of solids (such as silt, clay, or organic material), liquids, or gases.

Turbidity may be due to natural flow conditions which include fine-grained sediments, high natural flow rates, aqueous chemistry promoting colloidal stability, and geochemical transients due to waste disposal.[31] Usually, however, excessive turbidity results from an improperly constructed or developed well or from purging and sampling the well at a higher rate than for which it was developed.

Turbidity has been shown to be a more sensitive indicator of equilibrated conditions for purging a well than the more commonly used field parameters, which include temperature, specific conductance, pH, redox, or dissolved oxygen.[30] Field parameters such as pH, Eh, specific conductance, and dissolved oxygen appear to be relatively insensitive for indicating well stability for colloids or for low-solubility, particle-reactive contaminants such as polycyclic aromatic hydrocarbons (PAHs). Variations in concentrations of PAHs, which are expected to be associated with colloids in ground water, followed the variations in turbidity during pumping.[8]

Excessive turbidity in ground water samples is of concern due to the possibility of introducing nonrepresentative ions and metals into samples and adsorbing contaminants onto the suspended particles where they may not be subsequently analyzed. Turbidity should be measured with initial well development. Depending on how a well was constructed and developed (or sampled), wells which deliver samples with turbidity values over 5 to 10 nephelometric turbidity units (NTUs) may have to be replaced (Figure 4.8).[45]

The standard instrument for determining low turbidities measures the intensity of light scattered at right angles to the incident light. Turbidity meters may operate continuously or may require taking discrete aliquots (Figure 4.9). A photoelectric nephelometer automatically compares the amount of light scattered in the sample to a reference cell. A calibrated slit turbidimeter is operated manually by visually comparing a light source to the light passing through the sample.[7] To operate a slit turbidimeter the operator turns a dial controlling the slit opening until the field is of uniform brightness.

The sensitivity of a turbidimeter should permit detecting turbidity differences of 0.02 NTUs or less in waters having turbidity of less than 1 NTU, with a range of 0 to 40 NTUs. A tungsten-filament lamp operated at a temperature of between 2200 and 3000 dK has been recommended for use in turbidity meters.[1] The distance traveled by incident and scattered light in the sample should be no greater than 10 cm. For measuring turbidities over 40 NTUs, the sample should be diluted with one or more volumes of turbidity-free water until the turbidity is between 30 to 40 NTUs.

Turbidimeters should be operated in accordance with the manufacturer's operating instructions. In the absence of a precalibrated scale, calibration curves are prepared by using a reagent grade Formazin standard or styrene divinylbenzene beads. Formazin is a carcinogen, so it must be carefully

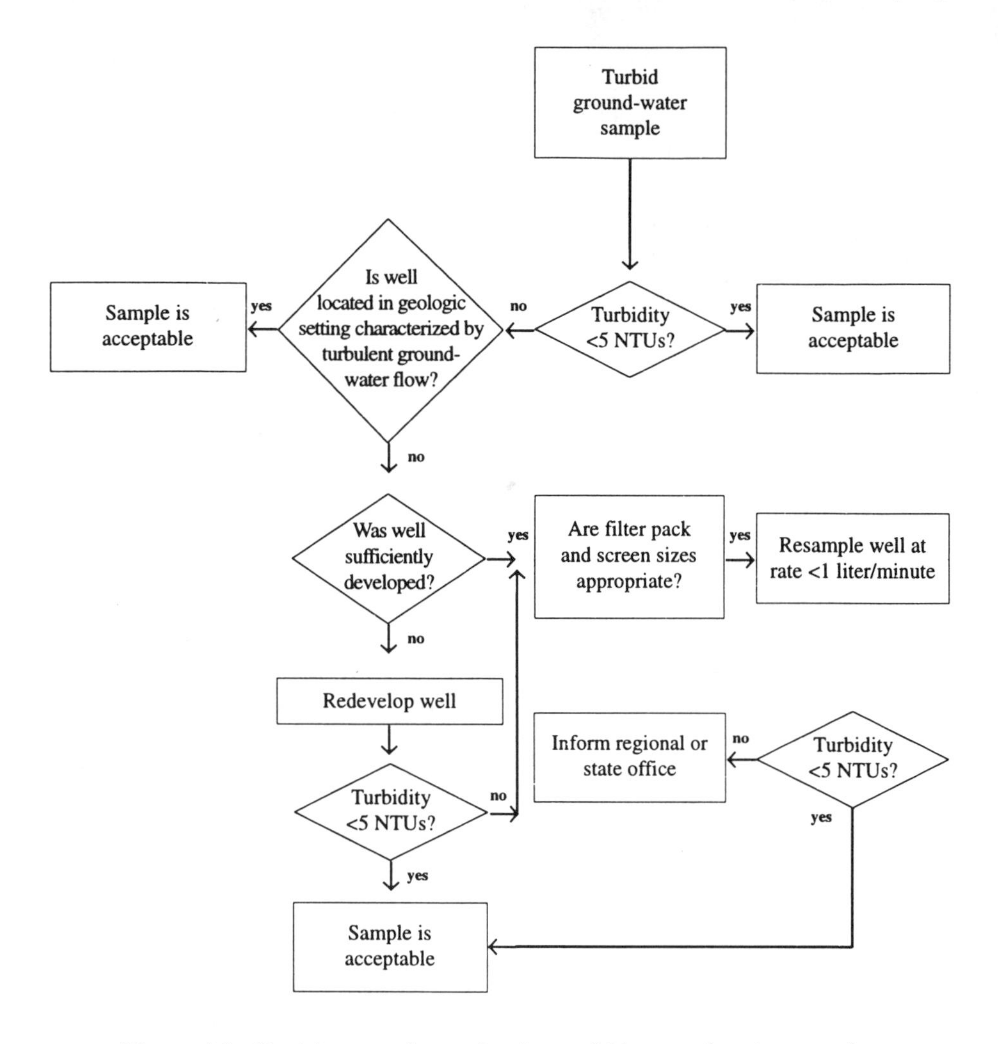

Figure 4.8 Decision tree for evaluating turbid ground water samples.

handled. Air bubbles, floating or suspended solids, and turbulent motions may cause false readings from a nephelometer. The sample tubes must not be handled where the light contacts them and must be kept scrupulously clean inside and out. Tubes should be discarded if they become scratched or etched.

Alkalinity

Alkalinity is defined as the capacity to neutralize an acid and is usually expressed as millequivalents, or milligrams per liter of $CaCO_3$. Alkalinity (acid neutralizing capacity) can be determined by titrating the sample with a strong acid to a preselected equivalent point.[38] Alkalinity in most natural waters is produced by the dissolved carbon dioxide species, bicarbonate and

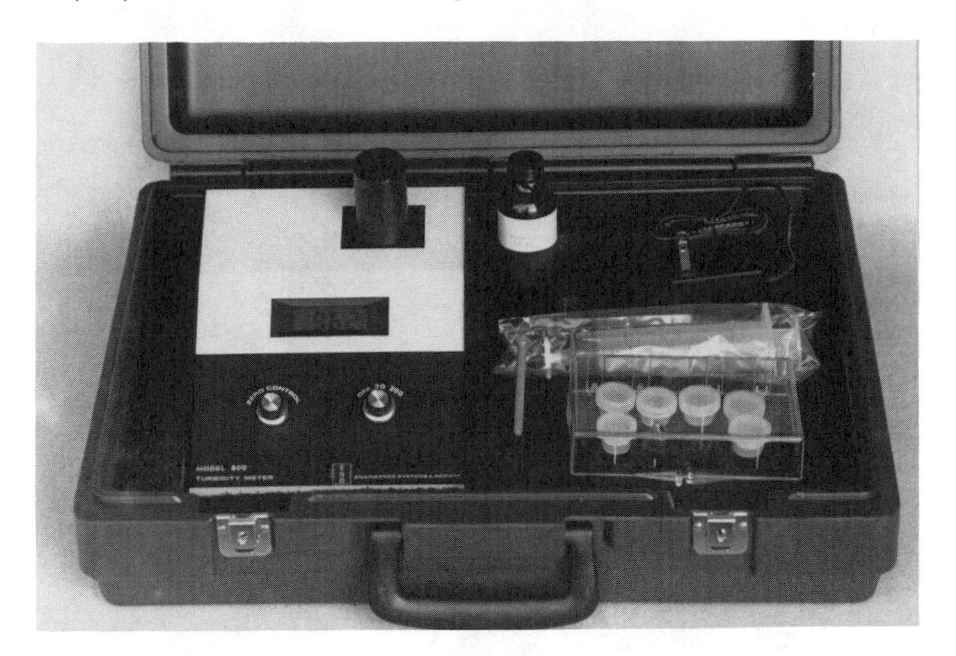

Figure 4.9 A discrete aliquot turbidity meter. (Photograph courtesy of Engineered Systems and Designs; *Newark, Delaware.*)

carbonate.[23] Alkalinity can only be interpreted in terms of specific substances when the chemical composition of the sample is known.[1]

Alkalinity is critical in establishing solubilities and toxicities of some metals and the buffering capacity of some waters.[7] In ground water systems, measurement of alkalinity can assist in determining the susceptibility to shifts in pH, which may relate to trace metal mobility.

Alkalinity should be determined in the field, as it may change rapidly after obtaining the sample. Alkalinity is not sensitive to gaseous carbon dioxide exchange, but is sensitive to changes in redox, hydrolysis, complexation, or precipitations of inorganic, organic, or biologic nature. Differences of 33 mg/L alkalinity as HCO_3 have been observed between alkalinity taken in the field and samples measured in a laboratory after 5 to 120 days.[36] If the field alkalinity determinations are of lesser precision than those obtained in a laboratory, they may still compensate for alkalinity changes that occur prior to analysis within the specified holding times.

Alkalinity can be determined by means of an electrometric titration (using a pH meter) or by colorimetric titration (Figure 9.10). If the sample is turbid or colored, the colorimetric method will not work and filtration of the sample to reduce turbidity for alkalinity determinations is not an approved method.[41]

Electrometric titration is a more precise and accurate method for determining alkalinity than the colorimetric method. The presence of suspended solids may interfere with this method. The electrometric method involves

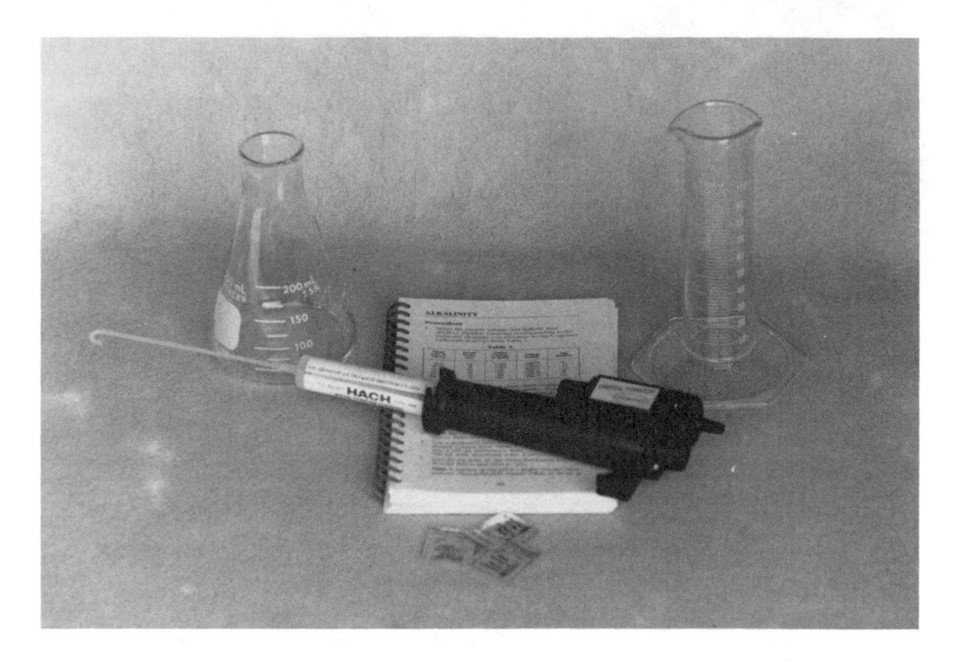

Figure 4.10 An alkalinity titration kit. (Photograph courtesy of Hach, Inc.; *Loveland, Colorado.*)

titrating 100 mL of the sample to a pH of (usually) 4.5 with 0.02 *N* reagent grade hydrochloric acid in 0.5 mL increments. With each 0.5 mL addition of acid, the pH is taken with a pH meter and recorded. A titration curve is then produced which defines inflection points and indicates buffering capacity.[7]

Colorimetric titration of a sample can be undertaken by using an indicator such as bromocresol green if the water is not naturally colored or turbid. The presence of waste material in the sample may destroy the indicator. The sample is titrated with an acid until the pH of the sample is 4.5.[42] Alkalinity can then be calculated by an equation, using the amount of acid required to change the pH to 4.5. When titrating, the sample must not be filtered, diluted, concentrated, or altered in any way.

Other measurements made in the field

Field measurements of ions such as chloride, sulfate, and nitrate with ion-specific electrodes may be inadvisable because interference from certain constituents make it difficult to calibrate the probes.[33] Kits which colorometrically provide concentrations of contaminants of concern in the field are becoming more common. These methods may or may not have USEPA approval as equivalent methods. In some cases, colorometric field analysis kits may provide useful information about whether a particular contaminant is or is not present in ground water.

References

1. APHA, *Standard Methods for the Examination of Water and Waste Water*, 18th ed., American Public Health Association, Washington, D.C., 1992.
2. ASTM, *Manual on Industrial Water and Industrial Waste Water*, American Society for Testing and Materials, Philadelphia, PA, 1962.
3. ASTM, D 1498-76, *Standard Practice for Oxidation-Reduction Potential in Water*, American Society for Testing and Materials, Philadelphia, PA, 1981.
4. ASTM, D 1125-82, *Standard Test Methods for Electrical Conductivity and Resistivity of Water*, American Society for Testing and Materials, Philadelphia, PA, 1983.
5. ASTM, D 1293-84, *Standard Test Methods for pH of Water*, American Society for Testing and Materials, Philadelphia, PA, 1985.
6. ASTM, D 4750-87, *Standard Test Method for Determining Subsurface Liquid Levels in a Borehole or Monitoring Well (Observation Well)*, American Society for Testing and Materials, Philadelphia, PA, 1988.
7. ASTM, D 1067-88, *Standard Test Methods for Acidity or Alkalinity of Water*, American Society for Testing and Materials, Philadelphia, PA, 1989.
8. Backhus, D.A., Ryan, J.N., Groher, D.M., MacFarlane, J.K., and Gschwend, P.M., Sampling colloids and colloid-associated contaminants in ground water, *Ground Water*, 31(3), 1993.
9. Barcelona, M.J., Gibb, J.P., Helfrich, J.A., and Garske, E.E., Practical Guide to Ground-Water Sampling, U.S. Environmental Protection Agency, Washington, D.C., 1986.
10. Barcelona, M.J., Holm, T.R., Schock, M.R., and George, G.K., Spatial and temporal gradients in aquifer oxidation-reduction conditions. *Water Resources Res.*, 25(5), 991, 1989.
11. Barcelona, M.J. and Holm, T.R., Oxidation-reduction capacities of aquifer solids, *Environ. Sci. Technol.*, 25(9), 1565, 1991.
12. Barcelona M.J., Wehrmann, H.A., and Varlijen, M.D., Reproducible well-purging proceedures and VOC stabilization criteria for ground-water sampling, *Ground Water*, 32(1), 1994.
13. Berner, R.A., A new geochemical classification of sedimentary environments, *J. Sedimentary Petrol.*, 51, 359, 1981.
14. Drever, J.I., *The Geochemistry of Natural Waters*, Prentice-Hall, Englewood Cliffs, NJ, 1982.
15. Edmunds, W.M., Miles, D.L., and Cook, J.M., *A Comparative Study of Sequential Redox Processes in Three British Aquifers, Hydrochemical Balances in Freshwater Systems*, edited by E. Eriksson, International Association of Hydrological Sciences, Velp, The Netherlands, 1984.
16. Essaid, H.I., Herkelrath, W.N., and Hess, K.M., Air, Oil, and Water Distributions at a Crude-Oil Spill Site, Bemidji, Minnesota, U.S.G.S. Toxic Substances Hydrology Program: Proceedings of the Technical Meeting, Monterey, CA, March, 1991.
17. Fetter, C.W., *Contaminant Hydrogeology*, Macmillan, New York, 1993.
18. Freeze, R.A. and Cherry, J.A., *Groundwater*, Prentice-Hall, Englewood Cliffs, NJ, 1979.
19. Garrels, R.M. and Christ, C.L., *Solutions, Minerals and Equilibria*, Freeman and Cooper, San Francisco, 1965.

20. Garske, E.E., and Schock, M.R., An inexpensive flow-through cell and measurement system for monitoring selected chemical parameters in ground water, *GWMR*, Summer, 1986.
21. Gray, D.M., Upgrade your pH measurements in high purity water, *Power*, March, 1985.
22. Heron, G., Christensen, T.H., and Tjell, J.C., Oxidation Capacity of Aquifer Sediments, *Environ Sci. Technol.*, 28(1), 1994.
23. Hem, J.D., Study and Interpretation of the Chemical Characteristics of Natural Waters, U.S. Geological Survey, Water Supply Paper 2254, U.S. Department of the Interior, Washington, D.C., 1985.
24. Hostettler, J.D., Electrode electrons, aqueous electrons, and redox potentials in natural waters, *Am. J. Sci.*, 284, 734, 1984.
25. Hudak, P.F., Clements, K.M., and Loaiciga, H.A., Water-table correction factors applied to gasoline contamination, *J. Environ. Eng.*, 119, 1993.
26. Lenhard, R.J. and Parker, J.C., Estimation of free hydrocarbon volume from fluid levels in monitoring wells, *Ground Water*, 1990.
27. Lindberg, R.D. and Runnells, D.D., Ground water redox reactions: an analysis of equilibrium state applied to Eh measurements and geochemical modeling, *Science*, 225, 925, 1984.
28. Liu, C.W. and Narasimhan, T.N., Redox-controlled multiple species reactive chemical transport. I. Model development, *Water Resources Res.*, 25(5), 869, 1989.
29. Plazak, D., Differences between water-level probes, *GWMR*, Winter, 1994.
30. Puls, R.W., Powel, R.M., Clark, D.A., and Paul, C.J., Facilitated Transport of Inorganic Contaminants in Ground Water: Part II. Colloidal Transport, EPA/600/M-91/040, U.S. Environmental Protection Agency, Washington, D.C., 1991
31. Puls, R.W. and Powel, R.M., Acquisition of representative ground water quality samples for metals, *GWMR*, Summer, 1992.
32. Richardson, S.M. and McSween, H.Y., *Geochemistry Pathways and Processes*, Prentice-Hall, Englewood Cliffs, NJ, 1989.
33. Ritchey, J.D., Electronic sensing devices used for in-situ ground water monitoring, *GWMR*, Spring, 1986.
34. Rose, S. and Long, A., Monitoring dissolved oxygen in ground water: some basic considerations, *GWMR*, Winter, 1988.
35. Schoenleber, J.R., Morton, P.S., Field Sampling Procedures Manual, State of New Jersey, Trenton, 1992.
36. Schock, M.R. and Schock, S., Effect of container type on pH and alkalinity stability, *Water Res.*, 16, 1455, 1982.
37. Stolzenburg, T.R. and Nichols, D.G., *Preliminary Results on Chemical Changes in Groundwater Samples Due to Sampling Devices*, Electric Power Research Institute, Palo Alto, CA, 1985.
38. Stumm, W. and Morgan, J.J., *Aquatic Chemistry*, Wiley-Interscience, New York, 1981.
39. Slabaugh, W.H. and Parsons, T.D., *General Chemistry*, John Wiley & Sons, New York, 1976.
40. Testa, S. and Pacskowski, M., Volume determination and recoverability of free hydrocarbon, *GWMR*, 9(1), 1989.
41. USEPA, Alkalinity Method 310.2 (Automated), U.S. Environmental Protection Agency, Washington, D.C., 1974.

42. USEPA, Alkalinity Method 310.1 (Titrimetric), U.S. Environmental Protection Agency, Washington, D.C., 1978.
43. USEPA, Methods for Chemical Analysis of Waters and Wastes, U.S. Environmental Protection Agency, Washington, D.C., 1979.
44. USEPA, RCRA Ground-Water Monitoring Technical Enforcement Guidance Document, U.S. Environmental Protection Agency, Washington, D.C., 1986.
45. USEPA, Chapter Eleven of SW-846 Ground Water Monitoring, Final Draft, U.S. Environmental Protection Agency, Washington, D.C., 1991.
46. Walton-Day, K., Macalady, D.L., Brooks, M.H., and Tate, V.T., Field Methods for Measurement of Ground Water Redox Chemical Parameters, *GWMR*, Fall, 1990.

chapter five

Equipment and techniques used for obtaining ground water samples

"Listen you rebels, must we bring you water
from this rock?" Then Moses raised his arm
and struck the rock twice with his staff.
Water gushed out, and the community and their livestock
drank. But the Lord said to Moses and Aaron,
"Because you did not trust in me enough to honor me
as holy in the sight of the Israelites, you will not bring
this community into the land I give them."
The Bible, Numbers, 20:10–12 (NIV)

Introduction

The equipment and techniques used for ground water sampling must be refined as more information becomes available on how to optimize ground water sampling. In the last 10 years numerous studies have been undertaken to determine what effects sampling equipment and techniques have on ground water samples. The verified and accepted results of ongoing basic research into ground water sampling need to be applied to routine ground water monitoring to ensure that the most representative ground water data are obtained.

Variations in ground water sampling and analysis from site to site include differences in hydrogeochemical setting, sampling equipment, sampling protocols, sampling crews, and variabilities inherent to analytic laboratories. Despite these variations, data from one site should be comparable to another qualitatively if not quantitatively.

This chapter is intended to provide an overview of sampling equipment and techniques and how to optimize ground water sampling plans and sampling proficiency. Adequate information is available to make inferences about ground water sampling equipment and techniques, but the best professional judgment still will be required to design site-specific sampling and analysis plans and ensure that sampling crews provide accurate ground water samples.

Vertical solute concentrations

Solute concentrations in ground water may vary in time and in three dimensions. Contaminant concentrations in ground water depend on whether the source of contamination has been reduced or eliminated and on the physical and chemical properties of the contaminant, attenuative and advective processes, and degree of anisotropy of the aquifer. Depending on the chemical, biological, and advective-dispersive processes in an aquifer, monitoring wells may only provide qualitative data on the solutes in ground water. Vertical integration of solutes along the length of a screen in a typical monitoring well using standard purging procedures may underestimate ground water contamination by orders of magnitude.[47]

How much vertical change in solute concentration may be occurring and how much vertical chemical resolution is necessary for achieving program goals are important questions to consider. Large vertical concentration gradients such as discrete zones of concentrated LNAPLs (light, nonaqueous phase liquids) or DNAPLs (dense, nonaqueous phase liquids), small seams of highly transmissive zones, or small layers of fine material in the aquifer which may pond DNAPLs above the rest of the water column are a few examples of how large vertical concentration gradients can exist within small vertical zones in the screened interval of a monitoring well.

If zones can be identified within an aquifer that are contributing a disproportionately high quantity of contaminants, then during well installation and subsequent sampling these zones can be targeted for sample acquisition. However, some wells intended to be used for compliance (detection) monitoring or for monitoring free phase liquids may need longer screens to provide better vertical coverage. Installing a well with a surface seal to approximately 20 ft and no annular grout seal for the remainder of the cased interval has been suggested in order to intercept contaminants in the vadose zone which may otherwise not be detected.[21]

Multiport samplers that sample discrete zones within a permanent installation can assist in providing vertical gradient resolution. Analytical costs may become prohibitively expensive to do much in the way of serious, ongoing vertical profiling. One detailed vertical chemical profile (e.g., centimeter intervals) may not suffice to adequately characterize an aquifer due to water level fluctuations and because concentrations change with time. Real-time on-site analysis in a mobile laboratory may help to resolve questions about vertical chemical gradients.

Natural fluctuations of the water table or alterations of water levels from inconsistent or unacceptable purging may also alter vertical chemical gradients. Consideration of potential vertical alterations caused by inconsistent or unacceptable purging and sampling further underscores the need to use a consistent technique from sampling event to sampling event. Until accurate analysis (DQO Level 3; see Chapter 7) of a large quantity of ground water samples becomes more cost effective, then some degree of vertical integration from monitoring wells may be necessary.

Using push-type samplers for optimal placement of permanent wells, closely examining good boring logs, geophysically logging existing wells, considering the physical and chemical properties of the contaminant, and vertically profiling the screened interval of the well (using temperature, pH, redox, dissolved oxygen, specific conductance) may assist in determining the number of samples required for analysis to characterize the site to meet data quality objectives and/or where to place a sampling device within the screened interval of a conventional monitoring well.

Ground water sample acquisition devices

Selection of ground water sampling equipment is based on the hydraulic characteristics of the aquifer, the diameter of the well, the sampling depth, equipment and personnel costs, and the chemical constituents in the ground water. Sampling equipment should be reliable, portable, durable, and easy to clean and operate. Sampling equipment should be inert with respect to the chemical parameters to be sampled and should minimize sample alterations including degassing, volatilization, sorption, and leaching.[6] Whenever a relatively sophisticated piece of equipment is purchased (either a pump or field parameter equipment), it may be a good idea to get formal training on the equipment by the manufacturer or sales representative.

The more commonly used ground water sample acquisition devices include:

- Grab mechanisms such as bailers and syringe devices
- Suction-lift mechanisms, which include peristaltic and centrifugal pumps
- Positive displacement mechanisms, which include gas-drive devices, bladder pumps, and electric submersible pumps
- Inertia pumps

If ground water contaminated with volatile organic compounds (VOCs) is to be sampled in a deep well that has been developed in a highly transmissive aquifer, different sampling equipment and techniques may be appropriate than those for sampling major anions and cations a shallow well developed in a formation with low permeability. For deeper wells, more pumping lift capacity is required to bring water to the surface, and some types of sampling equipment may interfere with chemical constituents more than other types. The following overview of sampling equipment is intended to provide a basis for evaluating the equipment and techniques used for ground water sampling.

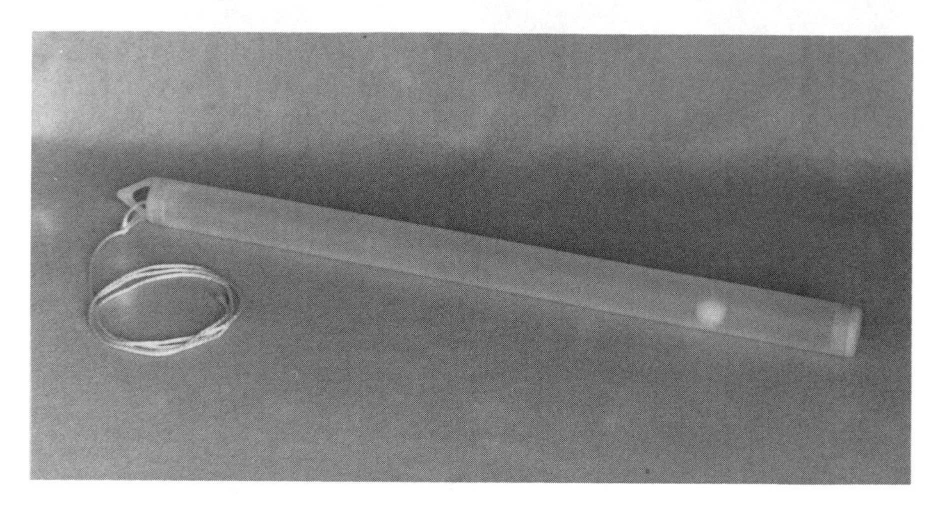

Figure 5.1 A single check valve Teflon™ bailer. The ball check valve is visible inside the bailer body, toward the bottom of the bailer.

Grab mechanisms

Bailers

The simplest and least expensive device commonly used to purge wells and collect ground water samples is the single check valve bailer (Figure 5.1). A bailer is basically a hollow tube with a ball check valve at the base. The ball check valve should have a specific gravity of between 1.4 and 2.0[1] and must allow the use of a bottom emptying device to reduce sample aeration.

Bailers are made of Teflon™, PVC, stainless steel, or polyethylene and come in various lengths and diameters. Bailers need to be laboratory cleaned, stored, and transported carefully wrapped to avoid contamination. Disposable bailers (one-time use) are also readily available. Field cleaning of bailers generally is not recommended.

A dual check valve bailer (also referred to as a point source bailer) is similar to a single check valve bailer with the exception that a second check valve is incorporated into the top of the bailer. The second check valve functions to reduce mixing of water within the bailer and the well casing as the bailer is withdrawn and to reduce the potential for introducing well artifacts or atmospheric gases into the sample. The inlet and outlet of point source bailers are sometimes tapered to allow water to move more freely through the unit.

New lengths of clean, strong, braided polypropylene line tied to the bailer with a bow-line knot or new lengths or thoroughly cleaned PTFE-coated metal cable should be supplied with each laboratory-cleaned or disposable bailer.

As appropriate, an approved safety and health plan must be prepared and used for all laboratory and field activities (see Chapter 6). Methanol is flammable, and this needs to be addressed in the safety and health plan. The laboratory procedure commonly used for cleaning both Teflon™ and stainless steel bailers is outlined below. Disposable gloves should be worn during bailer disassembly, cleaning, and reassembly.

1. Bailers are first disassembled (if possible).
2. Bailer pieces are then scrubbed using nonphosphate laboratory detergent, hot tap water, and a clean industrial strength cylindrical brush on the inside and a scratchy pad on the outside.
3. The bailers should then be rinsed with hot tap water, followed by three bailer volumes of deionized water.
4. In a well-ventilated area the bailer is then immersed in a graduated cylinder containing laboratory grade methanol. After approximately 20 bailers have been immersed, the methanol should be discarded and fresh methanol supplied.
5. Allow all bailer pieces to air dry for approximately 30 min. If methanol is still beaded on the bailer after the drying period, it is still contaminated. A stronger detergent such as Soft Scrub® should be used, and the above process repeated.
6. Do not bake bailers until all solvents are evaporated, as they are flammable. Teflon™ bailers should be dried at 100°C in an air-dry oven for 20 min and stainless steel bailers are dried at 200°C for about 20 min.
7. Reassemble the bailers after they have cooled, wearing a new pair of clean disposable gloves.
8. Cleaned bailers are then wrapped in aluminum foil (shiny side out), placed into a small-diameter (inert) plastic bag sheath, stapled shut, and labeled as to whether it is Teflon™ or stainless steel and the cleaning lot number.
9. One bailer per cleaned lot should have deionized water run though it and be analyzed for the most sensitive parameter for which it might be used in the field (for example, VOCs).

Newly installed monitoring wells should be tested with a "dummy" bailer (or pump) of the same outside diameter and slightly longer than the intended sampling device to ensure that the bailer (or pump) will not get stuck. This may be particularly important if the well is out of plumb. A bailer allows water to pass through it as it descends. Turbulence is imparted to the water in the bailer by the check valve and the sides of the bailer, so the sample experiences some degree of mixing. The USEPA[61] recommends that a bailer be submerged only to a depth necessary for filling, to ensure consistent samples.

For a bailer to properly function it must be carefully lowered to the top of the water column and, when contact with the water surface is established, allowed to slowly sink. This procedure avoids inducing a pressure wave in the well that could dislodge fines in a manner similar to the surging action used to develop monitoring wells. The bailer must also be removed slowly from the water to reduce surging. Some sampling protocols call for three bailer volumes of sample water to be rinsed through a bailer prior to sample acquisition.

Bottom-emptying devices must be used with bailers. A bottom-emptying device may allow air bubbles to be displaced upwards into the sample water inside the bailer, which causes aeration of the sample. An 8% lower concentration of purgeable organic compounds has been observed when samples were poured from the top of a point source bailer vs. using a bottom-emptying device.[17]

Light nonaqueous phase liquids (LNAPLs) and dense nonaqueous phase liquids (DNAPLs) should be collected prior to purging activities. If the thickness of a LNAPL is 2 ft or greater, a single-check-valve bailer is the equipment of choice.[61] The bailer should be carefully lowered to a depth less than the LNAPL/water interface, as previously determined with an interface meter. If the LNAPL is less than 2 ft thick and is greater than 25 ft below ground surface, a weighted top-filling bailer should be used. The best method for collecting DNAPLs is to carefully use a double-check-valve bailer.[61] Special health and safety precautions must be taken when LNAPLs and DNAPLs are anticipated.

Bailers are inexpensive, relatively easy to clean, and portable; a separate disposable or laboratory-cleaned bailer can be used for each well at virtually any depth. Bailer use may be facilitated by using a fishing line down-rigger attached to the well. If in-line filtration is employed, conventional bailers may perform accurately for acquisition of trace metals.[13]

The use of a bailer to purge a well can be laborious and time consuming and offers the greatest chance of disturbing the chemistry of the well of any of the commonly used evacuation procedures with the exception of an air lift pump. Results obtained from a bailer are extremely operator dependent and therefore quite variable. Repeated insertion and withdrawal of a bailer causes significant surging, mixing, and aeration, even when operated carefully.[45]

Inconsistent operator usage and excessive purging have resulted in turbidity of >100 NTUs from a bailer, compared to <5 NTUs with a low-flow-rate pump.[46] Bailers also may provide unrepresentative samples due to contact with the atmosphere during retrieval, or when transferring the sample into sample containers. The USEPA currently allows the use of bailers for purging and sampling monitoring wells.

Syringe sampling devices

Syringe samplers operate much like a medical syringe. The device is lowered to the desired sampling depth and the plunger is moved either mechanically or pneumatically to draw water into the sample tube. Samples obtained with syringe samplers are subject to only a slight negative pressure and do not come in contact with atmospheric gases. If glass-bodied syringes are used, the body may be inert with respect to most parameters, and the sample can be shipped in the syringe body. Filtration of samples is readily accommodated, and the apparatus is inexpensive and very portable. Syringe samplers

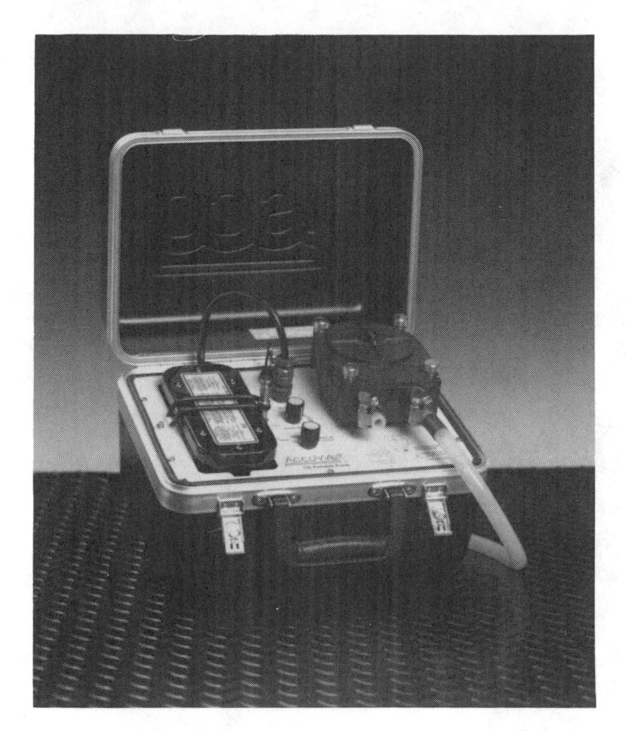

Figure 5.2 A peristaltic pump. (Photograph courtesy of Isco, Inc.; *Lincoln, Nebraska.*)

are limited in sample volume by the size of the syringe body. Some syringe samplers can retrieve a volume of up to 0.8 L. Suspended solids may cause leakage problems around the check valves and seals, exposing the sample to reduced pressures when the sampler is being filled or increased pressures when the sampler is being emptied.[24]

Suction lift mechanisms

Peristaltic pumps

A peristaltic pump is powered by an electric motor that rotates two or three pinch rollers in a circle, in a direction away from the water to be sampled (Figure 5.2). The rotation of the pinch rollers alternately squeezes and releases the flexible (usually silicone) tubing against the wall of the pump case, inducing a vacuum (Figure 5.3). Typically, the sample delivery tubing for a peristaltic pump is made of flexible PVC (e.g., Tygon™) tubing. Peristaltic pumps have a maximum operating depth of approximately 25 ft as they operate by inducing a vacuum. Peristaltic pumps are easy to transport, operate, and clean. Power can be supplied with rechargeable nicad batteries, or by using jumper cables from a vehicle's battery. Peristaltic pumps theoretically may cause degassing, by reducing the partial pressure of the sample. Degassing a water sample may cause loss of VOCs and accelerate CO_2 evolution, which consumes hydrogen ions and may result in an increase in pH.[31] Research comparing

Figure 5.3 The pinch roller operating mechanism of a peristaltic pump. (Photograph courtesy of Isco, Inc.; *Lincoln, Nebraska.*)

various sample acquisition devices including peristaltic pumps is provided below. A peristaltic pump may be used to "vacuum" up a LNAPL if it is less than 2 ft thick and less than 25 ft below ground surface.[61] VOCs should not be in contact with the pump's internal silicone tubing, as the tubing may adsorb VOCs. The use of a vacuum flask ahead of the silicone tubing or the use of Teflon®™-lined silicone tubing may reduce the problem of sorption. All suction lift pump tubing should be equipped with a foot check valve to prevent water from the pump and tubing from falling back into the well.

Centrifugal pumps

A centrifugal pump consists of an impeller in a housing, attached to the shaft of a motor. The sample tubing runs from the pump down to the surface of the water. Prior to operation, the pump must be primed by filling the impeller housing and sample tubing with water. As the impeller turns, a partial vacuum is created which lifts water out of the well. Pumping rates of between 5 to 40 gal/min are possible.[51] Introduction of water to prime the pump and sample tube may cause sample alterations, and the flow rates may be too high to obtain representative samples. The impeller and housing of a centrifugal pump need to be made of chemically inert materials if it is used for sample acquisition.

Positive displacement mechanisms

Bladder pumps

A bladder pump consists of a source of compressed gas, a gas flow controller, sample lines, and the pump (Figure 5.4). A bladder pump operates by alternately inflating and deflating a flexible bladder usually made of Teflon™. Water enters through a strainer or screen past the lower check valve into the bladder, either from the overlying hydrostatic pressure in the water column, forcing water into the pump during the refill cycle (when air pressure is vented), or from an elastic construction of the bladder. During the sample discharge cycle, compressed air is forced into the space between the bladder and the pump body, closing the bottom check valve and squeezing the bladder. This forces water upwards past the upper check valve, into the sample discharge line. The recharge-discharge cycle is repeated to bring water to the surface. Different manufacturers of bladder pumps may incorporate alternative designs into their pumps, but the operating principle generally is the same (Figure 5.5).

Bladder pumps operate by positive displacement (push the sample upward) and as such reduces the potential for VOC loss from an induced vacuum. The compressed air used to operate a bladder pump does not come into contact with the sample water.

Compressed air is delivered from the surface to the pump by means of either a cylinder of compressed gas or an air compressor and must be regulated by a flow controller that alternately allows inflation and deflation of the bladder. The pumping rate is controlled by adjusting the duration of the pumping and venting cycles with the flow controller. The maximum pumping rate depends on the particular pump and the height the pump must lift the water. Bladder pumps can be adjusted to accommodate higher rates of flow for purging and lower flow rates for sampling. Flow velocities of bladder pumps may be misleading, because actual flow at the surface is averaged over the two-stage flow cycle.[40]

Bladder (and submersible) pumps may be purchased with inflatable packers made of polyvinylchloride or Viton™ to isolate the pump from water above, below, or both above and below the pump intake. If a packer is used, it must be made of materials which are inert with respect to the target parameters.

The term "dedicated" has been used to describe equipment used only once per sampling event (e.g., a dedicated bailer) or sampling equipment that is permanently installed (e.g., a dedicated pump). The author has chosen the latter as the definition for dedicated equipment. Bladder pumps may be dedicated to each well or may be transported from well to well to purge and sample multiple wells.

Some bladder pumps require the hydrostatic pressure of the overlying water to operate them and may not operate if the pump is not deep enough into the water column. Bladder pumps have high sample contact surface

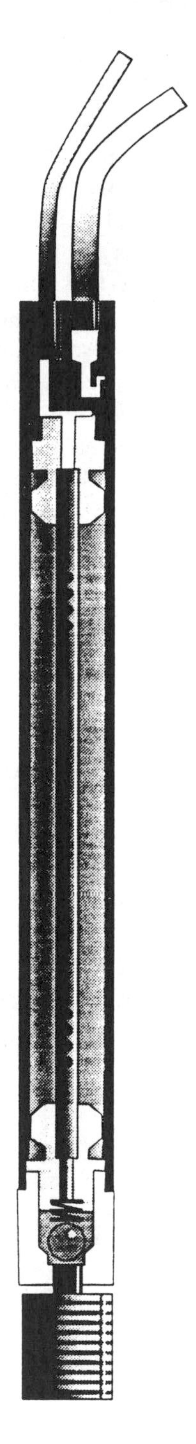

Figure 5.4 A diagram of a bladder pump (courtesy of Isco, Inc.; *Lincoln, Nebraska.*)

Figure 5.5 A complete bladder pump system with compressor, flow controller, tubing, and two pumps in the foreground. (Photograph courtesy of Isco, Inc.; *Lincoln, Nebraska.*)

areas constructed of materials that may be unsuitable for sampling hydrophobic organic compounds.[4]

Bladder pumps require a compressor or a cylinder of compressed gas so they are not as portable as some other sampling pumps (e.g., a peristaltic pump). If a large number of wells are to be sampled and/or the wells are relatively deep, then a compressor is usually required because a cylinder may not last long enough or be as cost effective. Gasoline-powered compressors (and generators) need to be located as far downwind from the sample discharge point as is possible to avoid contamination by gasoline or the exhaust fumes.

Gas displacement pumps

Gas displacement pumps operate much like a bladder pump but do not utilize a bladder. The compressed gas comes into contact with the water inside the pump reservoir and may alter the pH because of changes in carbon dioxide concentrations, so sample acquisition is not recommended with

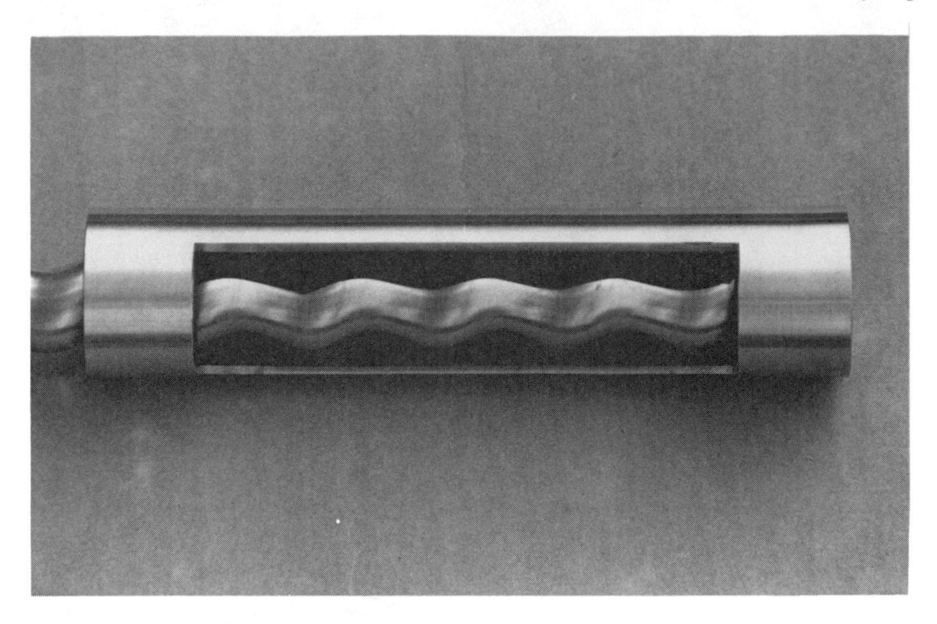

Figure 5.6 Cutaway of a helical rotor submersible pump. (Photograph courtesy of Keck, Inc.; *Williamston, Michigan.*)

these types of pumps even if nitrogen gas is used to operate the sampler. The proper timing of the pressurizing and venting cycles may be tedious, and they must be reset whenever the depth of pumping is changed.[30] A gas displacement pump can operate at a much higher pumping rate than a bladder pump. If a bladder pump is selected to sample a well, then the same compressed gas source and flow controller may be used to operate the gas displacement pump for purging the well.

Electric submersible pumps

Electric submersible pumps operate by using impellers, meshing gears, or a helical rotor (Figure 5.6). A centrifugal submersible pump utilizes an impeller or pump stages in series to positively displace water upwards (Figure 5.7). A gear submersible pump has a set of meshing Teflon™ gears (much like paddle wheels) that push the water along the pump walls, and the close tolerance of the gears prevents water from flowing backwards. A helical rotor submersible pump uses a spiral rotor that turns against the walls of the pump chamber, the stator, which is usually made of a semiflexible Viton™, to produce a progressive cavitation. The electric motor driving the submersible pumping mechanism is housed within the pump assembly. Depending on the pump type, a small gasoline-powered electric generator or an automobile battery is needed to supply power to the electric motor within the pump assembly (Figure 5.8).

A submersible pump operates by positive displacement and as such reduces the potential for VOC loss from inducing a negative pressure. Excessive turbulence may be imparted to the sample, especially at higher flow rates, and submersible pumps may generate excessive heat (especially at

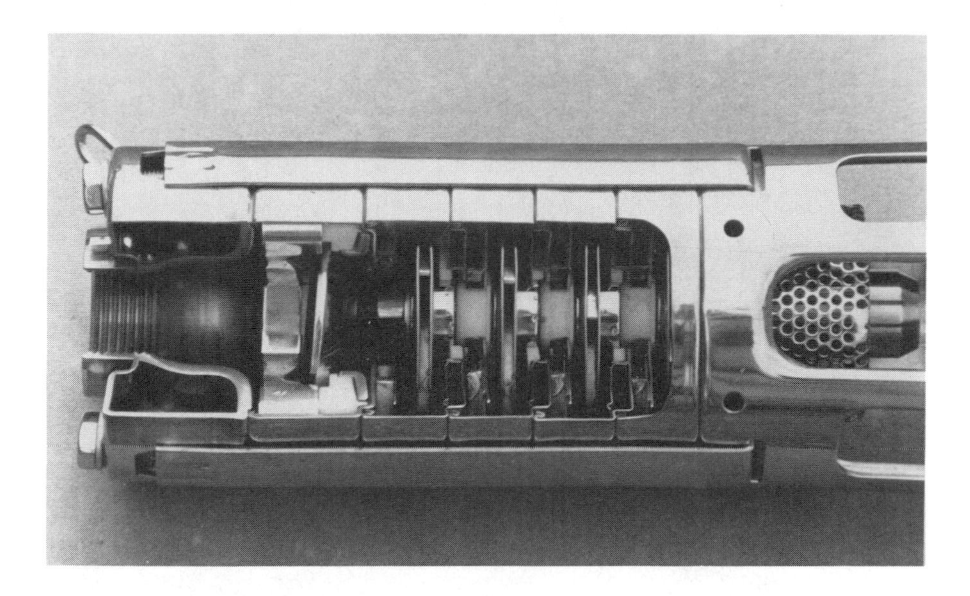

Figure 5.7 Cutaway of a centrifugal submersible pump.

lower flow rates) that could impact concentrations of VOCs.[40] Research comparing various sample acquisition devices including submersible pumps is provided below.

Electric submersible pumps may have only one operating flow rate, and this may be a problem if a lower purge or sample rate is required. Using a flow restrictor which "throttles down" by means of valving is not an acceptable means of reducing flow, as this may introduce the potential for partial pressure changes and subsequent chemical alterations of the samples.[20] Frequent stalling of some submersible pumps at low flow rates (<100 mL/min) has been reported. Sand may cause submersible pumps to "sand lock" (cease to operate due to sand jamming the pump).

Before using any pump, the operating instructions should be carefully read. When lowering a submersible pump the weight of the pump should be used to keep the power cord taut or the pump may become wedged between the power cord and the side of the well. The voltage and amperage should be monitored using a volt/amp meter when operating an electric submersible pump. A sudden increase in the current level used to operate the pump may indicate

1. A poor electrical connection or a cut in the cable
2. That total solids have increased
3. That the unit may be becoming sand-locked
4. The impellers may have swelled due to chemical interaction with the contaminated water, or
5. The discharge tubing is obstructed

Figure 5.8 A submersible pump and tubing reel without flow controller or electric generator. (Photograph courtesy of Instrumentation Northwest, Inc.; *Redmond, Washington.*)

Submersible pumps use the water as a coolant and must never be operated dry or they may overheat and burn out.

Air lift pumps

Air lift pumps bring water to the surface by introducing a compressed gas into the sample line via a small gas introduction line. The gas displaces the water, and the water is forced to the surface. This method bubbles gas directly into the sample and can greatly alter sample constituents. This method is not recommended for either purging or sample acquisition.

Jet pumps

A jet pump is a centrifugal pump that receives a boost from a nozzle-venturi arrangement. Centrifugal pumps create a lower pressure inside the pump than outside, and the nozzle-venturi also creates a low pressure. The combination of low pressures causes water to move toward the discharge point of the pump. The reduction in pressure in the nozzle-venturi may cause a change in partial pressure, resulting in an alteration of the sample.

Gas-operated piston pump

A gas-operated piston pump is a self-priming reciprocating piston-type submersible pump. The pump consists of two in-line water cylinders joined by an intermediate gas chamber and has a switching unit through which a common connecting rod passes to the pistons.[27] Gas in the intermediate switching unit alternately enters and exhausts without contacting the sample, driving both pistons either up or down. During the upstroke, water is forced out of the top chamber past the top check valve towards the surface; simultaneously water is drawn into the bottom cylinder. During the downstroke, water is drawn into the top cylinder and forced out of the bottom cylinder past the bottom check valve towards the surface. Flow rates of 0.25 to 0.5 gal/min at a depth of 500 ft are possible with these pumps. The flow rate, however, may only be varied over a limited range and may be unsuitable for reducing the flow between purging and sampling. Water with a high suspended solids content may cause problems with the operation of the pump. The complete pumping system may weigh several hundred pounds, limiting its portability in the field.

Inertia pumps

Inertia pumps consist of a tube or pipe with a check valve at the bottom. The pump operates by manually or mechanically raising and lowering the tube repeatedly, causing water to move upwards due to inertia. Manually operated inertia pumps are relatively inexpensive but may become tiring to operate. Some inertia pumps may rapidly lower the water level in wells developed in low permeable formations.

Experimental sampling equipment used in monitoring wells

When extremely low sensitivity (<10 mg/L) is not critical, analysis by gas-purging a ground water sample inside the modified syringe sampler used to obtain the sample allows twice as many samples to be analyzed at the same time as the more conventional purge-and-trap method. This method also enables the less volatile compound such as polynuclear aromatic hydrocarbons, phenol, and cresols to be determined.[39]

Another nascent sample collection technique exclusively for VOCs acquisition involves using adsorption/thermal desorption (ATD) with 60/80 mesh Tenax™, a sorbent (Figure 5.9). ATD involves allowing a known volume of water to pass through a sorbent cartridge at a fixed rate, either at the surface or in the screened interval of the monitoring well. After sampling, most of the water is removed, and the cartridge is thermally desorbed to a gas chromatographic column. This method appears to minimize volatilization losses from sampling, and lower detection limits (nanograms per liter) are possible than with the conventional purge and trap method. Statistically

Figure 5.9 An automated VOC sampler. (Photograph courtesy of Isco, Inc.; *Lincoln, Nebraska.*)

significant differences (lower ATD concentrations) however, were found between the ATD method and samples obtained using a pump and analyzed with the purge and trap method.[49]

A sampler equipped with a sorbent tube and a diffusional membrane impermeable to water but with a high permeability rate for organic vapors has been used in existing monitoring wells or may be driven into the sampling zone.[25] A time-integrated value may be realized, but the results need to be compared (calibrated) to known concentrations.

Other innovative sampling equipment is available for acquisition of ground water samples which reduces exposure of samples to the atmosphere and handling.

Pump tubing

Tubing used to transport the sample from the pumping device to the surface may be a potential source of error or bias. Tubing may sorb VOCs, may introduce a substantial amount of leachable organic matter, or may allow the diffusion of oxygen through the tubing, resulting in sample alterations.

Barcelona et al.[8] found that the relative affinity of tubing to sorption of chloroform, trichloroethylene, trichloroethane, and tetrachloroethylene was in the order of Teflon™ (PTFE) < polypropylene < polyethylene < flexible polyvinylchloride (e.g., Tygon™) < silicone. The bulk of the sorptive losses in closed lengths of tubing occurred in the first 20 min and desorption of the sorbed compounds took place almost immediately. Using 15 m of tubing with a 6-mm (1/4 in.) inside diameter, a 400-ppb mixture of halocarbons, and a flow rate of 100 mL/min, the predicted percent losses ranged from 11 for PTFE to 36 for silicone tubing. At constant flow rates, the predicted losses of halocarbons (comparing laboratory tests to field conditions) increase with larger diameter tubing.

At low flow rates, polymeric tubing may diffuse appreciable amounts of oxygen into ground water samples. The amount of oxygen transfer through tubing is proportional to tubing length and is inversely related to the pumping rate. In an experiment using PTFE tubing with a wall thickness of 0.0625 in., an outside diameter of 0.25 in., a flow rate of 100 mL/min, and a tubing length of 200 ft, approximately 0.6 mg/L of oxygen was diffused across the sample line. With a flow rate of 100 mL/min and a tubing length of 100 ft, approximately 0.25 mg/L of oxygen was diffused across the sample line.[22] Oxygen diffusion in PTFE sampling lines can introduce serious errors in ground water quality investigations, and should be minimized (see Figure 5.10).

More research into optimizing sampling tubing needs to be undertaken, and tubing which minimizes sorption/leaching and diffusion of oxygen should be used. If permanently installed equipment is used, then other tubing materials such as threaded, rigid PVC or stainless steel could be considered to reduce oxygen diffusion; however, leaching of metals such as chromium, nickel, molybdenum, iron, titanium, cobalt, and tungsten may occur if stainless steel is used.

The distance from the wellhead to the sample discharge point should be minimized, and the tubing must be protected from direct exposure to sunlight to reduce the potential for oxygen diffusion, photoxidation effects, and changes in temperature. Seasonal differences in air temperature caused the temperature readings in a flow-through cell to differ from the downhole probe value by 3 to 5°C, even though the sample discharge line from the wellhead to the cell was lined with foam tubing.[10] Sample tubing should be insulated against alterations from the ambient air temperature and shielded from exposure to direct sunlight to reduce photoxidation effects.

Depending on the site-specific sampling protocol, new lengths of tubing may be required for each well for each sampling event, tubing may be

Figure 5.10 Polyethylene tubing and submersible pump power cable on a reel.

dedicated to each well and be cleaned in a laboratory or in the field, or the tubing may be reused and cleaned in the field between wells. If a pump is used to purge a well followed by sample acquisition with a bailer, then at minimum the outside of the tubing should be rinsed with potable water. The equipment used, target parameters, and the use of equipment blanks (quality control samples) will dictate the type of tubing to use, and the cleaning procedure required.

Considerations for selecting sampling equipment and a sampling protocol

A sampling protocol should be based on the constituents to be analyzed, should ensure that the sample is representative of the formation, and should minimize alteration of the sample during the withdrawal process.[60] The decisions to be made for each site include:

- What sampling equipment will best minimize sample alterations
- If and how the well should be purged
- When a well can provide samples sufficiently representative of the formation
- What flow rates are appropriate
- If a sample should be field filtered, and if so how it should be done
- What field cleaning procedures are appropriate
- Which sampling equipment is the most cost effective

Each individual well in a network must be evaluated for the optimal sample acquisition equipment and technique. Dedicated pumps may not be optimal in all monitoring situations, such as very deep wells developed in a low recharging media such that the entire volume of water in the well annulus will be drawn up into the tubing during sampling. This scenario would dewater the filter pack, possibly even before completely purging the sampling lines.

Which sampling equipment will best minimize sample alterations

Numerous studies have been undertaken to evaluate the performance of ground water sampling equipment. Summaries of several research papers are provided to allow comparisons of ground water sample acquisition devices, as follows:

1. A low-speed submersible pump produced the least bias as compared to a bladder pump and a peristaltic pump when sampling ground water monitoring wells for trichloroethylene (TCE) and chromium. This was thought to be due in part to the rapid intake velocity of the bladder pump and degassing associated with peristaltic pumps. Peristaltic pumps recovered less TCE than a submersible or bladder pump and produced the highest dissolved oxygen, specific conductance, and pH.[40]
2. A 32-ft-long simulated monitoring well located in a stairwell was used to take samples simultaneously with acquisition devices. These samples were compared to samples taken at a stopcock at the base of the well. The losses of VOCs from the samples were determined to be predominantly from sorption onto the sampler and volatilization by exposure to bubbles, the head space in the sample container, or the atmosphere. A double-check-valve stainless steel and Teflon(TM) bailer, a syringe sampler, a prototype canister sampler, a bladder pump, a peristaltic pump, a double-valve sampler (gas displacement pump), and an inertia pump were tested using a simulated monitoring well spiked with five VOCs commonly occurring in ground water contaminant situations, each with differing tendencies to sorb or volatilize. Of 8 of the samplers, 7 produced samples with a mean concentration of 12% of the control data set. The Teflon(TM) bailer, double-valve sampler, and the bladder pump performed the best, based on statistically different mean concentrations.[3]

3. A total of 25 sets of samples was collected using a submersible pump, a Teflon™ bailer, and a bladder pump to be tested for nitrate and pesticides, including atrazine. Atrazine has a low vapor pressure and a low Henry's constant. It is less susceptible to volatilization and more prone to sorption losses than many common VOCs. Statistically there were no significant differences in concentrations among samplers.[52]
4. A total of 90 samples from 9 monitoring wells was sampled for VOCs ranging in concentration from 0.5 to 480 μg/L using disposable bailers and dedicated submersible pumps. Results indicated that there was no statistical difference between the two sampling devices.[38]
5. Mean concentrations of VOCs obtained from a piston pump and a centrifugal submersible pump were virtually identical. Laboratory analytical uncertainties from different analysts and analytical equipment appeared to influence the analytical results more than the pump type or sampling team.[28]
6. No difference was found in VOC sampling performance between a bailer and a home-made gas squeeze pump (a bladder pump) when sampling from a model monitoring well.[58]
7. At three sites (under field conditions) the best recovery of purgeable organic compounds (POCs) was (from highest to lowest): gear submersible pump, point source bailer, open bailer, helical rotor submersible pump, bladder pump, syringe sampler, and peristaltic pump. The overall standardized mean concentration of the first five samplers was closely grouped, suggesting a lack of any real differences among them in their ability to recover POCs sampled under the conditions present at the three sites. The three samplers with the highest precision were positive displacement pumping devices. The three samplers with the lowest precision were grab samplers, but the differences were less than 5% of POC concentrations from the most to the least precise of the devices tested.[24]
8. Purgeable organic compound recovery was 11% lower using a point source bailer than using a helical rotor submersible pump or a downhole isobaric sampler.[17]
9. VOC concentration variability is low for sampling VOCs regardless of the method used provided that careful and reproducible procedures are followed.[34]
10. A purging and sampling method using a bladder pump, submersible pump, jet-pump, bailer, air-lift pump, and inertia pump had only a few, and usually very small (a maximum of 4 mg/L of sodium, the most conservative cation), effects on inorganic water chemistry.[23]
11. Centrifugal, submersible pumps operated at a purge rate of 15 to 53 L/min (a high flow rate) caused agitation of the sample and the screened zone. Centrifugal, submersible pumps produced samples with statistically higher metals concentrations than bladder pumps operated at a purge rate of 0.5 to 4 L/min.[12]

12. At three geologically different sites with shallow wells, a peristaltic pump consistently produced the most reproducible results for metals. In wells deeper than 30 ft, a bladder pump gave the most reproducible results.[42]
13. Significant differences existed in particle population size between a low-speed bladder and submersible pump, and a high-speed submersible pump. Increasing the pumping rate generally increased turbidity and brought larger particles into suspension. Equilibrated turbidity with a peristaltic pump (at the Elizabeth City site) was generally less than 2 NTUs, while with a bailer it was greater than 200 NTUs.[45]
14. Bailed samples at a coal-tar contaminated site contained 10 to 100 times greater colloid concentrations and up to 750 times greater polycyclic aromatic hydrocarbons than samples taken with a submersible pump using a low flow rate. An enormous difference in the size of the particles between a pumped (generally less than 5 μm) and bailed samples (1 to 100 μm) also was observed. The pumped sample was composed predominantly of clay particles, and the bailed sample of silicon, presumably quartz, grains.[4]
15. Differences in the concentration of constituents during and after purging may be attributable to compositing samples from the entire length of the screen (instead of the type of sampling device or method employed), depending on the vertical distribution of solutes and the amount of drawdown which occurs during purging and sampling.[32]

A discussion of these abstracts is incorporated into a broader summary at the end of this chapter.

If and how the well should be purged

The following techniques are currently being used for acquiring ground water samples:

- Purge and sample with a bailer
- Purge with a pump and sample with a bailer
- Purge and sample with the same pump
- Use a packer to minimize purge volume and sample with a pump
- Use dedicated equipment to purge and sample or sample without purging

The question of how to approach well purging is critical to obtaining representative samples. Ground water in the upper part of the casing is typically not representative of the water in the aquifer because stagnant water is not free to interact with formation water. Also, the stagnant water is in contact with the casing and well gases for extended periods of time. Stagnant casing water and formation water in the screened interval usually have different temperatures, pHs, redox potentials, total suspended and dissolved solids, and different chemical compositions due to precipitation of metals, effervescence of dissolved gases, and loss of VOCs.

Errors introduced through improper purging of stagnant water were found to be greater than errors associated with sampling mechanisms, tubing, and construction materials.[7] The most common problem with purging is overpumping the formation.[20] Overpumping at a high rate may (1) increase turbidity, (2) cause dilution which could mask the presence of contaminants, or (3) dehydrate a portion of the saturated zone which could aerate and degas the formation water, causing unacceptable alterations in the chemistry of the water.

Several possible strategies are used to separate the overlying stagnant water from the water to be sampled. A commonly used technique involves removing the column of stagnant water in the casing (purging) from the top or bottom of the stagnant water column. Other techniques used with or without purging include an inflatable packer to isolate the sampling zone or dedicated equipment located within the screened interval or lowering a sampling device below the stagnant zone into the screened interval.

If dedicated equipment is not selected, then in medium- to high-yielding wells that are shallow, or wells that are screened at or near the water table, the USEPA[61] has advocated purging by placing the pump at the air/water interface. For medium- to high-yielding wells screened considerably below the static water level, the USEPA has advocated placing the pump within the screened interval and operating it at a moderate rate.

However, pumping from immediately beneath the top of the water column may reduce the volume of water which needs to be removed above the screen to two or three bore volumes, compared to possibly over ten times as much if the pump is placed substantially below the air-water interface.[48] A pump test and/or a packer may be used to reduce purge volumes. The USEPA has advocated purging low-yielding wells to dryness, unless by so doing water will cascade down the screen, thus aerating the sample. If cascading is a concern, then the evacuation should be carried out slowly.

A model of a monitoring well, consisting of a well screen and riser pipe placed in a container of water, was used in conjunction with a fluorescent dye introduced into the "stagnant zone" to simulate various purging methods. It was determined that an average of 2 to 4% of the water pumped from locations above the screen and an average of 1% of the water pumped from within the screen came from stagnant water above the pump inlet.[58]

A down-hole camera attached to peristaltic pump tubing indicated that placement of the camera itself created the greatest turbidity, with a gradual decrease in colloidal density over time while pumping at a rate of 0.2 L/min. In the absence of pumping overnight, re-equilibration after insertion of the pump was required in order to achieve steady-state colloidal density in the screened interval.[45]

A colloidal borescope consisting of a charge-coupled-device camera, optical magnification lens, and an illumination source has been used to assess the stability of colloids after insertion of equipment into the well.[26] During installation of the borescope in a variety of hydrogeologic settings, a massive disturbance of the flow field has been observed. After a period of time, ranging from a few to approximately 30 min, laminar flow in the same direction as local ground water flow replaces turbulent flow. Maximum colloidal density is observed upon insertion of the borescope in the well, and colloids appear to be stabilized within 24 hours of insertion of the borescope. The significant disturbances of colloidal density following pump emplacement seem to indicate that dedicated equipment will greatly reduce colloidal disturbances and increase sample representativeness.

When a well can provide samples representative of the formation

Although ground water flow rates form more of a continuum, two broad categories of hydrogeologic settings can be considered in monitoring situations: monitoring wells developed in relatively coarse-grained, rapidly recovering hydrogeologic media, and monitoring wells developed in relatively fine-grained, slowly recovering hydrogeologic media. Different sampling strategies for these two broad conditions are warranted, as they have their own separate constraints. Whichever purge and sampling strategy is proposed to be employed, approval by regulatory agencies based on an evaluation of the proposed methodology may be required.

When rapidly recovering wells can be said to be stabilized

Most wells with the screen set below the water table and developed in media where the ground water flow rate is relatively rapid have two distinctly different chemical zones present between water in the screened interval and water in the casing, with a sharp break occurring on either side of the beginning of the screen. This is not ubiquitous, as differences in well construction, hydrogeologic media, and horizontal and vertical ground water flow from well to well may cause a more gradual chemical gradation between the casing and screen.

Wells developed in rapidly recovering media have been said to be representative of formation water based on:

A specified number of well volumes
Stabilization based on field parameters
Minimal (micro) purging
Hydraulic characteristics of the individual well
Time series sampling

These criteria are summarized as follows.

A specified number of well volumes. In terms of purging criteria, one purge volume has been defined as:

1. The volume of water in the screened interval
2. The volume of stagnant water overlying the top of the screened interval
3. The volume of water in the screened interval, plus the overlying stagnant water
4. The pore volume of water in the filter pack, screened interval, plus the overlying stagnant water (pore volume plus casing volume)

In more transmissive aquifers, water in the screened interval of a monitoring well is free to mix with water in the aquifer, so a purge volume may be defined as the volume of stagnant water above the screened interval. It is a common (and relatively easy) practice for field personnel to calculate one bore volume to mean the total volume of water in the well. The volume of water in the screen and casing is calculated by the following formula:

$$V = [(DB) - (DW)] \times [(\pi)(r^2)] / [231]$$

Where

V	=	volume (in gallons) of water in the screen and casing
DW	=	measured depth to water from the measuring point of the monitoring well, including the stick-up (in inches)
DB	=	measured depth to bottom of the well, including the stick-up
r	=	radius of well (in inches)

In a study by Robbins and Martin-Hayden,[47] the curve that appeared to calibrate the best for calculating optimal purge volumes for VOCs was the one that defined the well volume as the pore volume in the saturated sand pack plus the volume of water in the casing. The USEPA[61] has recommended that three to ten casing and filter-pack volumes be removed for purging, and the actual number should be determined on a site-specific basis. Several authors have suggested that numbers of bore volumes, ranging from less than 1 to more than 20, be removed prior to taking ground water samples. Gibb et al.[15] concluded that removing four to six bore volumes is sufficient in most situations to produce representative samples, but should be confirmed by using a pumping test and verification by field parameters. Three casing volumes appear to be reasonably reliable for stability of field parameters but only works about half the time for purgeable organic compound (POC) concentrations.[16]

Stabilization based on field parameters. Purging a monitoring well based on stabilization of field parameters such as temperature, specific conductance, and pH has been undertaken by measuring and recording the field parameters in the purge water after each bore volume has been removed. In the past, well stability was indicated if two or three successive bore volumes were stabilized with the specified field parameter criteria. This was typically undertaken with much higher purge and sample rates than 1 L/min for purging and 100 mL/min for sampling, so purging and sampling could be undertaken relatively rapidly. If a minimal purge strategy is selected with dedicated equipment, using 1 L/min for purging, then taking field parameters more frequently (e.g., every half bore volume) to determine well stability prior to sampling could be considered. At flow rates of ~1 L/min oxygen, conductance, and VOC levels stabilized consistently after pumping less than one-half bore volume.[10] When field parameters have stabilized within the following ranges over successive (potentially partial) bore volumes, stable sample chemistry is indicated and sampling may commence:

pH:	±0.1 pH units
Specific conductance:	±10.0 μS/cm
Temperature:	±0.5°C
Dissolved oxygen:	±0.2 mg O_2
Turbidity:	10 NTUs or less, ±2 NTUs (the turbidity stabilization criteria has yet to be formally established)

Dissolved oxygen and specific conductance were found to be the most useful field indicators for stabilization of VOCs during low flow rate purging, as pH and temperature readings stabilized almost immediately.[10] Specific conductance, pH, and temperature were found to be the least sensitive of the indicator parameters to indicate well stabilization for metals; contaminant concentrations, dissolved oxygen, redox, and turbidity are the most sensitive.[46] Specific conductance may best reflect major ion chemistry in ground water. This is because specific conductance is less affected by either volatilization or small-scale heterogeneity.[10] Turbidity has been shown to be a more sensitive indicator of equilibrated conditions for metals when purging a well than the more commonly used field parameters,[44] and turbidity should be used to determine when to collect a sample.[4] An example of a field stabilization form is provided in Chapter 6. Stabilization of field parameters, however, may not accurately represent when a well should be sampled for VOCs.[16] Field parameters are not ideally suited to monitor changes in POCs, because they are not chemically related to POC concentrations and some field parameters are not chemically conservative during sampling. Chemical partitioning and biological degradation between various POCs in the ground water may cause the number of purged volumes for a stable value to differ. POC concentrations stabilized when three casing volumes were purged in only 55% of the cases in the study, and POCs did not stabilize at the same times as the measured field parameter. By contrast, the stabilized number of casing volumes determined by UV absorbence at 254 nm using 0.45-μm field-filtered samples measured with an ultraviolet-visible spectrophotometer were very close to those of the aromatic compounds when the total aromatic concentrations exceeded 150 μg/L.[16] Stabilization of indicator parameters using dedicated pumps at fixed rates of pumping (~1 L/min) may, however, provide consistent results.[10]

Minimal purging. If the pump is properly placed and operated at a low flow rate within the screened interval, then most purging may be unnecessary. If samples are taken too quickly after insertion of the sampling device into the well, if the pumping rate is too high, if a bailer is used, or if the pump is not below the casing-screen interface, then the chemistry of the screened interval chemistry may be biased. High-rate purging and sampling may (1) increase the potential for sorption-desorption reactions, (2) mix water from multiple aquifer zones, (3) involve large investments of time, and (4) may involve handling large volumes of purge water which may be hazardous waste. Based on the disappearance of a latex colloid from the screened interval, the time required to allow a monitoring well screened in medium sand and clayey sand to recover from tubing or pump installation can be as high as 120 hours.[44] Using sodium bromide, sodium chloride, and deionized water tracers in six monitoring wells installed in fine to coarse sand, it was determined that within 24 hours of tracer introduction water in the screened interval returned to background concentration. Water above the screened interval remained at the initial concentration throughout the test. The use of permanently installed (dedicated) sampling devices can be used to obtain a representative sample, without purging the well.[48] Some purging may still be required when using permanently installed dedicated sampling equipment and low flow rates. Even in the type of transmissive hydrogeologic environment where it would be expected that ground water in the screened interval would be renewed by flow, differences of 15 to 23% lower values between initial (after removing the "dead" volume in the pump and lines) and final (postpurging, using field parameters for stabilization) VOC values have been observed, and these differences are modest compared to other results.[10] In wells with low VOC concentrations, differences between the initial and final sample may be negligible.

If horizontal ground water flow is dominant in the screened interval (no vertical advection into the overlying stagnant water), then diffusion would be a primary mechanism to remove gases from solution. Diffusion rate calculations indicate that exchange of gases between the atmosphere and the ground water are not a concern.[26] If a minimal purge strategy is used, some stagnant water must still be removed from the sampling lines and pump prior to sample acquisition, and stabilization needs to be verified by field parameters.

The hydraulic and chemical characteristics of an individual well. These characteristics of an individual well should be established prior to sampling and monitored for the useful lifetime of the well.[37] A 2- to 3-hour pumping test, simultaneously taking field parameters and time series sampling for target parameters after the well has been initially developed and at specific time intervals afterwards, would assist in determining an optimal purge volume and rate for the specific well and to detect any changes over time. Well

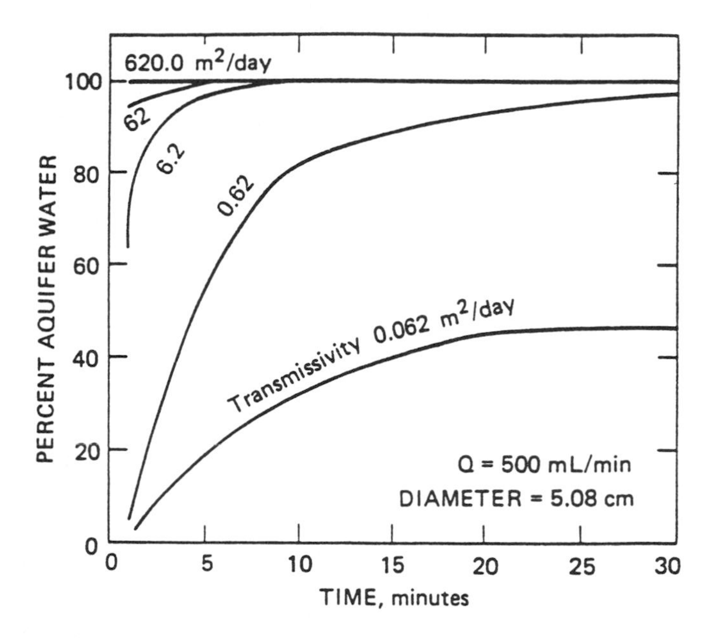

Figure 5.11 Percentage of aquifer water vs. time for different transmissivities. (Adapted from Barcelona, M. J. et al., *Practical Guide to Groundwater Sampling,* Illinois State Geological Survey Contract Report 374, Springfield, 1986.)

purging estimates based on hydraulic performance are undertaken by first determining hydraulic conductivity, using either a pump or slug test. The percentages of aquifer water pumped for a 2-in.-diameter well for a range of transmissivities have been calculated and are provided in Figure 5.11. The following example is provided to show how to estimate purge volume based on hydraulic conductivity data.

Example: Monitoring well MW-5 was properly installed and developed and was subsequently slug tested. A purge rate of 500 mL/min and a sampling rate of 100 mL/min have been suggested to minimize turbidity. If the pump intake is placed at the top of the screen then (A) what percentage of aquifer water will be pumped after 5 min, and (B) how much water should be pumped to approximate 100% aquifer water?

Given:

MW-5 is a 2-in.-diameter well, and is 30 ft deep
The screen is 5 ft long
The aquifer is 15 ft thick
The static water level is 5 ft below land surface
The hydraulic conductivity is 10^{-3} cm/sec.

Solution A:

1. Aquifer transmissivity (T) = hydraulic conductivity ×aquifer thickness
   10^{-3} cm/sec = 10^{-5} m/sec
   15 ft = 4.57 m
   [10^{-5}] ×[(4.57)] = 0.0000457 m^2/sec; 86,400 sec/day;
   T = [0.0000457] ×[86,400] = 3.948 m^2/day
2. Using Figure 5.11, after 5 min of pumping at 500 mL/min approximately 90% of the water is aquifer water

Solution B:

1. First calculate the volume of water in the screen and casing:
   [(30) – (5)] ×[(0.613 L/ft for a 2-in.-diameter well)] = 15.3 L
2. Next calculate the number of well volumes (N) after pumping 5 min:
   N = [(5 min)(0.5 L/min)]/[(15.3L)] = 0.163 well volumes, = 2.49 L

From this example, it can be seen that within 5 min of pumping a relatively high percentage of water is aquifer water if the pump intake is placed at the top of the screened interval. Due to potentially unrealistic assumptions made with some aquifer solutions and because hydraulic parameters may change over time, using the hydraulic performance of the well to calculate purge volume provides only an estimate of the purge volume required to stabilize a well.

Time series sampling. Time series sampling entails taking field parameters and collecting and analyzing samples at one, two, four, six, and ten bore volumes twice a year for the first three years.[29] Although the resolution on purge volume vs. ground water chemistry may be excellent, the cost of taking and analyzing samples using this method may be prohibitively expensive. Sampling less costly but analogous parameters or using less precise and less costly field analytical methods could be considered for time series sampling, to determine an optimal purge strategy.

When a slowly recovering well can be said to be stabilized

For slowly recovering wells several strategies have been proposed to obtain representative samples. These strategies include (1) not purging the well,[18] (2) purging the well and collecting the sample during recovery,[6,57] or (3) purging the well and waiting until complete recovery.[14] Monitoring wells where ground water flow is extremely slow or nondirectional may not be suitable for unpurged sampling due to the extremely slow volume turnover in the screen.[41]

Herzog et al.[19] found that VOC concentrations were significantly lower before purging than after purging in monitoring wells developed in low permeability materials and that at the 95% confidence interval there were no significant differences in chemical compositions from 2 to 48 hours after purging.

If the screen is set below the water table, the well should not be purged such that the water level falls below the screen, or else dewatering/aeration of the filter pack or water cascading down the sides of the screen may occur, both of which can readily air-strip VOCs. Volatilization losses of up to 70% within 1 hour after purging were reported in a laboratory test if the water level was allowed to drop below the screen.[33]

In low-permeability media with a small gradient it may take weeks or months for ground water to horizontally flow 2 in., which is the diameter of many monitoring wells. The slow movement of ground water through the screened interval of a monitoring well installed in low-permeability materials may cause the more volatile organic compounds to be advected and diffused to the air-water interface and lost to the atmosphere by volatilization.

Using a glass standpipe to simulate exposure of a monitoring well to the atmosphere, it was found that concentrations of trichloromethane, trichloroethylene, 1,1,1-trichloroethane, and tetrachloroethylene followed a first-order exponential decay, such that losses were about 10% within 12 hours and 99% within about a month.[33] If the well screen is set below the water table, then purging the well of the water above the screen and allowing some recovery (several hours) could be envisioned; this would remove a portion of the volatilized stagnant water and reduce cascading effects.

In fine-grained formations if a monitoring well is screened across the water table the two apparent choices are to not purge the well and lose VOCs through a stagnant water column between sampling events or purge the well and lose VOCs from cascading effects. At no time should a well be pumped to dryness if the recharge rate causes the formation water to vigorously cascade down the sides of the screen and cause an accelerated loss of VOCs.[60]

For acquisition of VOCs the first sample obtained from slowly recovering wells should be tested for field parameters and then collected in the order of volatilization sensitivity. The well should then have the field parameters retested to ascertain the stability of the water samples over time, whenever possible.

What purge and sampling flow rates are appropriate

In the past, a purge and sample rate of 500 and 100 mL/min, respectively, have been recommended for ground water sampling.[6] These pumping rates, however, may substantially increase the time (and associated costs) required to sample a well. Using low pumping rates may also increase oxygen diffusion through the tubing, and the temperature change from *in situ* to ambient air temperature may also alter the sample.

In a comparative study using a submersible pump vs. a bladder pump, the submersible pump (which pumped at a higher rate than the bladder pump) had considerably higher turbidity and metals concentrations than the lower flow rate bladder pump.[12] Low flow rate (0.2 to 0.3 L/min) purging and sampling have been demonstrated repeatedly to show no significant differences between filtered and unfiltered metal samples.[46]

A purge rate of 1 L/min and a sample rate of 100 mL/min have recently been suggested,[10] and these rates may represent a reasonable compromise to reduce purge time and sample alterations. The purge rate should never exceed the pumping rate used to develop the well, the sampling rate should not exceed the flow rate used for purging, and well purging should not exceed the well's recovery rate.[61] Using 200 mL/min to reduce the time to fill 40-mL purge and trap bottles (filling a bottle every 12 seconds) may be considered for inclement sampling conditions (windy, dusty, or rainy days).

Field filtering

As related in Chapter 4, total dissolved solids (TDS) have been defined as that portion of a sample that passes through a 1.5 μm filter, is then put in an oven at 180°C, and is subsequently weighed.[63] Ground water samples acquired for metals analysis routinely have been filtered in the field with a 0.45-μm filter to exclude portions of the sample that have been considered to be relatively immobile in ground water flow regimes. Samples obtained for VOC analysis generally are not filtered, as most volatile priority pollutants have a low to moderate affinity for solid substrates and VOC sample handling must be minimized.

The colloid state is a two-phase system in which one phase in a finely divided state is dispersed through a second phase. Colloids have been defined as particles with a diameter no larger than 10 μm,[54] and in ground water mobile colloids are generally smaller than 1 μm. Colloids in ground water include microorganisms, inorganic precipitates (hydrous iron, aluminum, and manganese oxides), humic material, and rock and mineral fragments, which include secondary clay minerals. Colloids may form from the dissolution of secondary minerals (iron oxyhydroxides and calcium carbonate) due to changes in ground water chemistry derived from ground water contamination.

Under certain hydrogeochemical conditions, the transport of inorganic colloids may be significant. Colloidal transport is dependant on ionic strength and composition, flow velocity, quantity, nature, and size of suspended colloids, geologic composition and structure, and ground water chemistry. In a column experiment, it was found that ionic composition of the supporting electrolyte and particle size were the most significant of these factors on colloidal transport.[64] Colloids can strongly sorb contaminants and have been shown to be transported hundreds of meters from their source. Colloidal particles in the range of 0.1 to 1.0 μm may be the most mobile in a sandy, porous medium,[36] and colloids larger than several microns in diameter are probably not mobile due to gravitational settling.[62]

The USEPA banned field filtering of ground water samples (*U.S. Federal Register*, 1991) based on research indicating that colloidal transport of hydrophobic and particle-reactive contaminants (polycyclic aromatic hydrocarbons, polychlorinated biphenyls, and metals including lead, chromium, copper, and nickel) was occurring in ground water and that field filtering may exclude part of the colloidal fraction that is naturally mobile.

Controversy surrounded the decision to ban field filtering due to concerns that field filtering may be appropriate in some instances and that not enough data exist to support the ban: "While there have been indications of this mechanism of (colloidal) contaminant transport, its significance remains a question."[45] The respective arguments of whether or not to field filter are presented and then discussed.

Why samples should be filtered

1. Filtering of ground water samples needs to be undertaken in wells that are inherently prone to excessive turbidity in order to exclude constituents that are not dissolved and hence not mobile. If total recoverable metals analysis of unfiltered samples is undertaken, the analysis may be more indicative of naturally occurring formation minerals or well artifacts than of contamination transport that may be occurring.
2. Contaminants such as PCBs, polynuclear aromatic hydrocarbons, phthalates, esters, and many pesticides are only slightly soluble in water. Laboratories typically extract these types of chemical constituents with an organic solvent in the laboratory, and if samples are unfiltered the compounds will desorb and appear as if they were in solution.[11]
3. Filtering needs to be undertaken to reduce the potential of clogging sensitive laboratory analytical equipment.
4. Acid preservation of turbid samples may cause dissolution of immobile particulates and so cause an overestimation of mobile metals concentrations.
5. Monitoring wells may become more or less siltier over time, and this variability may impart false trends into the data.

In summary, unfiltered metals samples make it difficult to distinguish truly mobile constituents from naturally occurring aquifer or well artifact constituents.

Why samples should not be filtered

1. If a ground water sample is anoxic and reduced and the pH of the water is greater than 7.4, dissolved ferrous iron oxygenation is autocatalyzed such that precipitation of amorphous ferric iron hydroxides can occur within seconds after initial aeration.[55] Iron precipitation can result in changes in pH, alkalinity, conductance, ionic strength, turbidity, and color.[35] Amorphous iron hydroxide will also adsorb trace metals. Filtering of ground water samples may physically or chemically alter the sample when transferring the sample to a filter, may cause oxygenation and subsequent alterations within the filter

substrate, or may cause alterations of dissolved gases if a vacuum is applied, which may alter pH and redox potential. These alterations of the sample may make the sample unrepresentative of the aquifer water.

2. Filtering may exclude some of the total mobile load of colloids and low-solubility contaminants,[4] resulting in an underestimation of contaminant load. If colloidal material of less than 0.45 µm in size passes through a 0.45-µm filter, then an overestimation of the dissolved concentration will result.[43]
3. Especially in karst terrains, field filtering should not be undertaken because colloid transport is far more likely to occur in conduit flow where colloid particles move more easily than in the larger solution channels.[61]
4. Filters typically become increasingly clogged during use, which reduces the nominal pore size and hence the cutoff for dissolved vs. particulate matter.
5. Variability is introduced by using different filter types, sizes, and filtering technique.

In summary, a 0.45-µm filter is an artificial convention, and the filtration step can cause as much sample disturbance as the sampling device itself.[53]

Discussion on field filtering

Aeration is considered to be the principal cause for chemical changes (in metals) during ground water sample collection.[13] Aeration becomes very important when the ground water is anoxic and reduced. Introduction of even a small amount of oxygen in anoxic and reduced ground water can result in decreases of up to 100% of lead, cadmium, zinc, arsenic, vanadium, phosphate, and possibly other trace metals. The amount of adsorption of trace metals onto ferric hydroxide depends on initial iron concentration, initial Eh and pH, redox and pH buffering capacity, and the extent of iron oxidation from sampling and sample handling including filtration.

In situations such as a properly designed and constructed monitoring well installed in well-sorted uniform sand, turbidity may be inherently low and field filtering is probably not needed if using low-flow-rate purging and sampling. Filtering of ground water samples must not be used to compensate for poorly designed or poorly constructed monitoring wells with excessive turbidity problems or poor sampling methodology.

Ground water sampling devices that cause the least disturbance (manifested as turbidity) also produce the most reproducible samples irrespective of filtration. Equilibrated turbidity units using a peristaltic pump have been generally less than 5 NTUs, but those obtained with a bailer have ranged from 5 to greater than 200 NTUs.[45] If the pH is less than 7 and a relatively large redox buffering capacity exists in the solution, then aeration effects of field filtering will not be as pronounced.[53]

Comparison of particle distributions of samples obtained from a bladder pump and a low-speed submersible pump to a high-speed submersible pump showed that the two low-speed pumps produced monomodal particle distributions of about 0.5 μm in size, whereas the high rate pump produced a bimodal distribution of slightly more particles with large sizes.[45]

Different conclusions have been drawn as to whether pronounced differences between filtered and unfiltered samples exist. The following conclusions may not agree as a result of using different equipment and techniques and/or differing physical or chemical characteristics of the aquifers and ground waters studied.

- When sampling technique is controlled, differences between filtered and unfiltered metals samples using 0.45-μm filters are not statistically significant, with the exception of iron.[12]
- The use of low-flow-rate purging and sampling with minimal disturbance of the stagnant water column consistently produced samples with no statistical difference between filtered and unfiltered samples, even in fine-textured glacial till.[46]
- Research at three different contamination sites using a 0.45-μm filter has shown that removal of potentially mobile colloids did not occur when using pumping rates of ~0.2–0.3 L/min, and that there were no significant differences in chromium concentrations between unfiltered and 5.0, 0.4, and 0.1 μm filtered samples.[45] Pumping rates should not greatly exceed ground water flow velocities or large differences between filtered and unfiltered samples may occur.
- A greater than 10% difference between filtered and unfiltered samples is typical for many elements. Differences in lead concentrations ranging from 20 to 600 times greater in the unfiltered vs. filtered samples and 6 to 24 times in chromium for unfiltered vs. filtered samples have been observed.[43]

The USEPA[60] has recommended taking one unfiltered sample for total metals and one filtered sample for dissolved metals. Puls and Barcelona[43] have recommended no filtration or a 5-μm filter for determination of mobile metals and in-line filtration with a large nonmetallic (e.g., 142 mm) polycarbonate-type 0.1 μm pore size filter for geochemical speciation models (i.e., the "dissolved" fraction). Taking two metals samples, however, doubles the already expensive (>$500/sample) analytical costs, and a filter with an effective pore size of 0.1 μm may clog up very quickly. With turbid samples Puls and Powell[46] advocate the use of a 5-μm filter where conditions are such that the acquisition of low-turbidity samples is not feasible due to geologic constraints, and that this condition can be documented. Use of dedicated equipment at low flow rates in properly constructed monitoring wells should obviate the need for field filtering in most situations. If field filtering is still seen as desirable, a 5-mm in-line filter should be considered, but regulatory approval may be required.

In the limited number of legitimate cases where suspended solids can be demonstrated to be inherently elevated, taking two samples (one filtered and one not filtered) should be considered. For regulatory programs, the unfiltered sample should be analyzed and if standards are exceeded then the filtered sample should be analyzed to determine which portions of the violations are attributable to the dissolved and exchangeable fractions.[56]

On May 1, 1992, a U.S. federal appeals court overturned a decision by the USEPA to put a landfill in Delaware on the National Priorities List. The decision was overturned because the USEPA acted "arbitrarily and capriciously" in taking only unfiltered ground water samples, contrary to the USEPA's own sampling guidance documents. To filter or not will depend on program requirements and monitoring objectives. The issue apparently has yet be settled, but meeting the test of hostile scrutiny in a court of law may need to be considered when deciding to filter or not to filter.

Colloidal characteristics such as size, elemental composition, solubility in the ground water, sorptive character, and surface charge can be assayed in ground water samples. These properties assist in determining where the colloidal material is coming from (e.g., a naturally mobile colloid or a well artifact), as well as other information such as the proportions of dissolved and colloidal phases and their relative mobility in the aquifer. Colloids can be collected in the field on a 0.015- to 0.03-μm membrane filter and then analyzed by a scanning electron microscope and energy dispersive X-ray diffraction.[4]

Field filtering procedure

For purposes of historical review or for ongoing field filtering, a discussion on filtering techniques is warranted. Three main types of filtration apparatus have been used in the field: in-line filtration, positive pressure filtration, and vacuum filtration devices (Figure 5.12). In-line filtration devices include one-use disposable cartridges and holders which house disposable filters.

In-line filtration is the preferred method of field filtration,[43] but the commonly used transfer vessel technique can yield precise results with the careful use of a bailer, peristaltic pump, or bladder pump.[13,53] Reproducible results from field filtering depend on sampling expertise, and a disposable in-line filter may be the simplest method to use without introducing errors. In-line filters should be pre-wetted with sample water and the outside of the cartridge should be tapped by hand to remove air bubbles trapped within the filter material prior to sample acquisition.

Most vacuum filtration devices require that the sample be transferred from the sampling equipment to the filtering device, and a vacuum is applied to pull the sample through the filter. When transferring a sample from the sampling device to the filtering equipment, the sample can be aerated, which can cause major changes in samples trace metal chemistry. Filtering must be undertaken as rapidly as possible after acquiring the sample to reduce con-

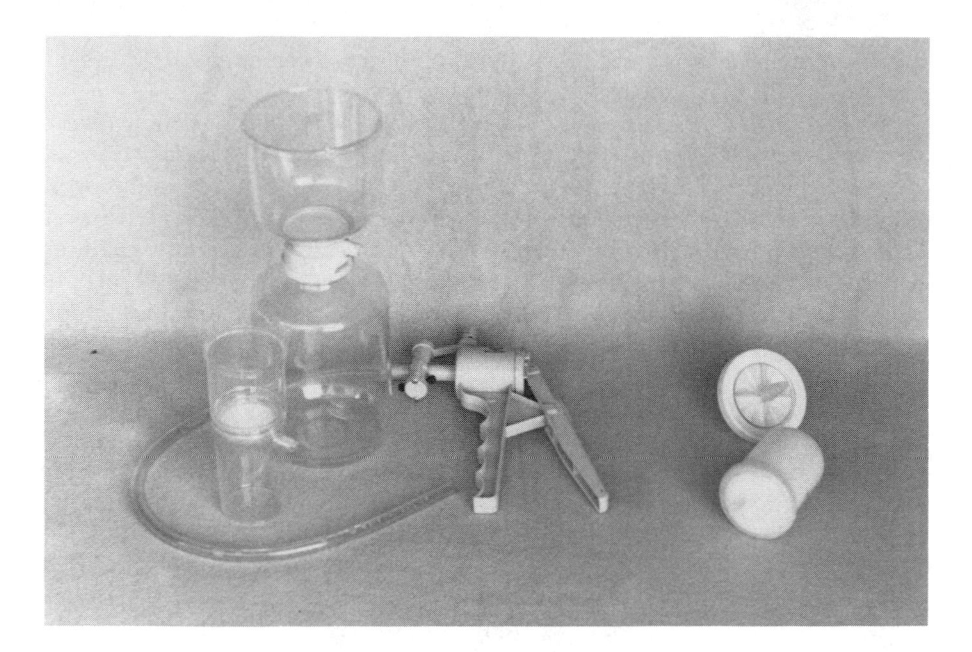

Figure 5.12 Field filters for ground water sampling; the two on the left are vacuum types and the two on the right are in-line types.

tact time with the air and subsequent iron precipitation. This is especially true when the dissolved oxygen and redox buffering capacity are low.

The outlet of the discharge tube of the low-flow-rate positive-pressure pump should be held below the sample water surface in the transfer vessel in order to keep aeration at a minimum. Use of a nitrogen-filled glove box to transfer samples into the filtering device has resulted in significant differences (>10%) between samples transferred in air and in nitrogen; this effect was more pronounced in wells with high iron content and low pH.[43]

Bailers are available that can pressure-filter the sample without removing the sample from the bailer. If the filtration equipment is not disposable it must be cleaned thoroughly between uses. Some membrane filters are hydrophobic and must be pre-wetted with propanol prior to use. Samples taken for metals analysis are put into polyethylene containers to reduce the potential for metals leaching from glass containers and immediately acidified to a pH of 2 with nitric acid (HNO_3).

Field cleaning equipment

Two separate procedures may be appropriate for decontaminating sample-contacting and nonsample–contacting equipment.[2] Cross contamination (carrying contaminants from one well to another) must be avoided. Some state programs do not allow the field cleaning of pumps or bailers. Cleaning the pumps in the laboratory and using new tubing has been suggested for each sampling event for each well.[65]

Bailers should be cleaned in a laboratory and used for only one well per sampling event. If a braided polypropylene line is used, it should be transported to the site in a clean plastic bag and kept from contaminants in the vehicle and field. The line should be used only once and then be discarded.

A pump used only for purging a well requires less scrupulous cleaning than one selected for purging and sampling. Depending on the objectives and relative costs, ensuring that the nondedicated sampling tubing is acceptably clean can be accomplished by one of the following:

- Use new lengths of tubing for each well and/or for each sampling event
- Use tubing for each well that can be laboratory cleaned and reused in subsequent sampling events
- Field clean the sampling line between wells

A sampling line can be expensive, but so are the costs of cleaning the line in a laboratory or in the field, and analysis of the required equipment blanks to prove the line is acceptably clean.

Field cleaning of equipment may require large quantities of potable and deionized water. The first step in field cleaning the sampling pumps and tubing is to clean the outside of the pump and tubing with a stiff brush and laboratory-grade nonphosphate detergent. A steam cleaner, hot pressure washer, or even a new (never used for pesticide application) hand pressurized "bug sprayer" should be used on the outside and on the inside, where possible. Next, using a new plastic overpack drum or garbage can (to avoid electric shocks with submersible pumps) a minimum of three equipment volumes of soapy, potable water should be pumped through the pump and tubing. The USEPA[61] recommends that sampling equipment used for acquiring organic constituents be rinsed with:

- Tap water
- Organic-free reagent water
- Reagent grade acetone
- Pesticide quality hexane, methyl alcohol, or isopropanol, depending on analyses

The USEPA[61] recommends that cleaning equipment used for sampling inorganic constituents should be washed with a nonphosphate detergent, and rinsed with:

- Dilute (0.1 *N*) hydrochloric or nitric acid
- Tap water
- Reagent water

When cleaning stainless steel in accordance with the safety and health plan, hydrochloric acid is preferred because nitric acid may oxidize the metal. The rinsate from the cleaning operation may need to be held awaiting analysis and possibly treated and disposed of as a hazardous waste. Field cleaning of equipment with just clean, soapy water may be possible, but as with any field cleaning protocol a suitable number of equipment blanks consisting of deionized water run through the equipment needs to be collected and analyzed.

A new pair of disposable gloves should be worn when cleaning the equipment. If the equipment is not to be used immediately then it should be allowed to dry thoroughly in a dust-free environment and packaged in oven-baked aluminum foil with a clean plastic bag over the outside. A label should be affixed to the outside wrapping stating the type of decontamination used and the date of decontamination. A custody seal may also be required on the cleaned equipment (see Chapter 6).

Which sampling equipment is most cost effective

When choosing a sampling protocol, the two constraints under which a manager is usually operating are (1) reducing the chance for sample alterations and (2) reducing costs. The major costs of sampling ground water monitoring wells are capital costs (equipment), supplies, and labor. Capital costs are considerably less for a bailer than a dedicated or nondedicated pump system. A new, reusable Teflon™ bailer (~$100) costs about 1/40th of a new bladder pump, compressor, and flow controller. Labor costs are increased with decreasing ease of access to the wells and decreased portability of the equipment and depend on the amount of field work required (setting up the equipment, ensuring stabilization, performing titrations, etc.) and the amount of field cleaning required.

A dedicated sampling system typically consists of a pump (including wiring or pneumatic lines) and sample delivery tubing. A portable gasoline-powered generator or an automobile battery is required to operate a submersible pump, and a gasoline-powered compressor or cylinder of compressed gas is required to operate a bladder pump.

Positive aspects of dedicated equipment include:

- Labor costs for setup and equipment decontamination may be significantly reduced
- If the pump is placed within the well screen, purging may be significantly reduced
- Using dedicated pumping equipment appears to greatly reduce turbidity and so provides a more reproducible sample
- The potential for cross-contamination between wells is reduced or eliminated

- Systematic errors introduced by sampling personnel are reduced with dedicated equipment, because fewer steps are required and the steps themselves are simpler

Negative aspects of dedicated equipment include:

- The equipment may initially cost more
- Equipment blanks may be more difficult to obtain
- Cleaning the pump and changing the sampling lines between sampling events may be required, depending on the potential for adsorption and subsequent leaching of contaminants

A well-prepared team of three individuals usually can sample 4 to 6 monitoring wells (5 to 75 ft deep) using nondedicated equipment in a full 8-hour day, exclusive of travel time.[6] It has been estimated that approximately $100 per well per sampling event was saved by using dedicated submersible pumps instead of disposable bailers at a Superfund site in California.[38] In general, a minimum savings of 25% per well per sampling event should be realized by using dedicated vs. nondedicated equipment.

The savings realized at a particular site by using dedicated equipment can be estimated by factoring in the respective labor and capital costs. The cost of a permanently installed bladder pump or low-flow submersible pump is approximately $500 to $1,000 per well, which is roughly the cost of analyzing one ground water sample for VOCs, trace metals, and anions and cations. When all the costs are included (not considering the costs associated with false data) permanently installed pumps are well advised.

If adsorption and subsequent leaching is possible in dedicated pumps and/or sampling lines, an equipment blank should be taken periodically by pulling the pump and lines of one of the most contaminated wells at the site and running laboratory-grade deionized water through the system at the same rate and volume as if a sample were to be taken. This may be more of a concern when monitoring dense, nonaqueous phase liquids, which have a proclivity to stick to pumps and tubing. In physically or chemically aggressive monitoring situations (highly corrosive, high concentrations of organic solvents, high total suspended solids, etc.), dedicated equipment may not be suitable.

If the equipment shows cross contamination from sampling event to sampling event within the same well, then between events the pump may have to be laboratory cleaned and the lines laboratory cleaned or replaced. The dedicated equipment should be returned to the well(s) within approximately 48 hours prior to sampling to reduce turbidity concerns. This increases labor costs but optimizes benefits realized from using dedicated equipment to reduce turbidity and mixing in monitoring wells.

A documented, site-specific periodic inspection and maintenance program for dedicated sample delivery equipment, including taking quality control samples from some of the wells in the network, needs to be considered. At minimum, this should be undertaken in at least one well in the network every other year and possibly more frequently in more monitoring points, depending on site specifics and data quality objectives.

Summary and review

Each individual well in a network must be evaluated for the optimal sample acquisition equipment and technique.

Selecting the optimal sampling equipment and techniques may seem complicated and sometimes contradictory, but some generalizations can be made to assist in preparing site- (and well-) specific sampling plans:

1. Accurate ground water sampling is highly operator dependent. Using equipment that most reduces sampling variability imposed by sampling crews is desirable.
2. Low-flow-rate purging and sampling is preferable to high-rate purging and sampling.
3. Bailers should be discouraged for sampling constituents other than nonaqueous phase contaminants. Bailers appear to be the device most prone to increasing turbidity by imparting a surging action and are the least preferable sampling device for trace metals in inherently turbid ground water flow regimes. If used very carefully by skilled operators, bailers may be able to produce VOC samples comparable with other types of sampling equipment.
4. Bailers may be used to carefully collect NAPLs and DNAPLs. The ground water should then be sampled using other sampling equipment. This method may make dedicated sampling equipment for NAPL and DNAPL acquisition more difficult.
5. Purging with a close-fitting (within the well annulus), dedicated, positive displacement pump positioned within the screened interval, at a rate of approximately 1 L/min, and sampling with the same pump at a rate of 100 mL/min subsequent to field parameter stabilization, appears to be *at this time* among the most scientifically defensible and potentially cost effective ways of acquiring representative ground water samples in many, if not most, monitoring situations. This method minimizes disturbances of the stagnant water column and reduces the potential for increasing turbidity. It reduces the chance for operator error(s), is less time consuming than using nondedicated equipment, and reduces the volume of potentially contaminated purge water. Dedicated pumps may not be optimal in all monitoring situations, such as very deep wells developed in a low recharding media with a small volume of water in the screen and casing.
6. When field parameters have stabilized within the following ranges over successive (whole or partial) bore volumes, stable sample chemistry is indicated, and sampling may commence:

pH:	±0.1 pH units
Specific conductance:	±10.0 μS/cm
Temperature:	±0.5°C
Dissolved oxygen:	±0.2 mg O_2
Turbidity:	10 NTUs or less, ±2 NTU (the turbidity stabilization criteria have yet to be formally established)

7. If dedicated equipment is not selected, then in medium- to high-yielding wells that are shallow or wells that are screened at the water table, purging should be undertaken by placing the pump at the air/water interface. For medium- to high-yielding wells screened considerably below the water level, the pump should be placed in the screened interval.[61] The sampling device should be placed at the desired sampling point in the screened interval, the stagnant water column above the screened interval should be minimally disturbed, and the sampling zone may be isolated with a packer; low flow rates (<1 L/min for purging and 100 mL/min for sampling) should be used; field parameters should be used to determine when sampling should commence; and tubing wall thickness should be maximized to exclude atmospheric gases.[45,46]
8. For wells developed in low-permeability materials, some purging appears to be appropriate. At no time should a well be purged to dryness if the recharge rate causes the formation water to vigorously cascade down the sides of the screen.[61]

References

1. ASTM, D 4448-85a, *Standard Guide for Sampling Groundwater Monitoring Wells*, American Society for Testing and Materials, Philadelphia, PA, 1986.
2. ASTM, D 5088-90, *Standard Practice for Decontamination of Field Equipment Used at Nonradioactive Waste Sites*, American Society for Testing and Materials, Philadelphia, PA, 1990.
3. Baerg, D.F., R.C. Starr, D.J.A. Smyth, and J.A. Cherry, Performance Testing of Conventional and Innovative Downhole Samplers and Pumps for VOCs in a Laboratory Monitoring Well, National Ground Water Sampling Symposium Proceedings, Washington, D.C., November 30, 1992.
4. Backhus, D.A., J.N. Ryan, D.M. Groher, J.K. MacFarlane, and P.M. Gschwend, Sampling colloids and colloid-associated contaminants in ground water, *Ground Water*, 31(3), 446–479, 1993.
5. Barcelona, M.J. and J.P. Gibb, Development of effective ground-water sampling protocols, *Ground-Water Contamination: Field Methods*, ASTM Spec. Tech. Publ. 963, American Society for Testing and Materials, Philadelphia, PA, 1988.
6. Barcelona, M.J., J.P. Gibb, J.A. Helfrich, and E.E. Garske, Practical Guide to Groundwater Sampling, Illinois State Geological Survey Contract Report 374, Springfield, 1986.

7. Barcelona, M.J. and J.A. Helfrich, Well construction and purging effects on ground-water samples, *Environ. Sci. Technol.,* 20(11), 1179, 1986.
8. Barcelona, M.J., J.A. Helfrich, and E.E. Garske, Sampling tubing effects on groundwater samples, *Analytical Chemistry,* 57(2), 460 1985 (correction in 57(13), 2752, 1985).
9. Barcelona, M.J., J.A. Helfrich, and E.E. Garske, Verification of sampling methods and selection of materials for ground-water contaminant studies, *Ground-Water Contamination: Field Methods,* ASTM Spec. Tech. Publ. 963, American Society for Testing and Materials, Philadelphia, PA, 1988.
10. Barcelona M.J., H.A. Wehrmann, and M.D. Varlijen, Reproducible well- purging procedures and VOC stabilization criteria for ground-water sampling, *Ground Water,* 32(1), 12–22, 1994.
11. Braids, O.C., *The Rationale For Filtration of Ground Water Samples,* Proceedings of the Sixth National Symposium and Exposition on Aquifer Restoration and Ground Water Monitoring, May 19-22, 1986, National Water Well Association, Dublin, OH, 1986.
12. Clark, S.B., N.M. Park, and R.C. Tuckfield, Effects of Sample Collection Device and Filter Pore Size on Concentrations of Metals in Groundwater Samples, National Ground Water Sampling Symposium Proceedings, November 30, 1992.
13. EPRI, *Preliminary Results on Chemical Changes in Groundwater Samples Due to Sampling Devices,* EPRI EA-4118, Project 2485-7, Electric Power Research Institute, Palo Alto, CA, 1985.
14. Fenn, D., E. Cocozza, J. Isbiter, O. Braids, B. Yare, and P. Roux, Procedures Manual for Ground Water Monitoring at Solid Waste Disposal Facilities, EPA/530/SW-611, U.S. Environmental Protection Agency, Washington, D.C., 1977.
15. Gibb, J.P., R.M. Schuller, and R.A. Griffin, Procedures for the Collection of Representative Water Quality Data from Monitoring Wells, Cooperative Groundwater Report 7, Illinois State Water and Geologic Surveys, Springfield, 1981.
16. Gibs, J., and T.E. Imbrigiotta, Well-purging criteria for sampling purgeable organic compounds, *Ground Water,* 28(1), 68, 1990.
17. Gibs, J., T.E. Imbrigiotta, Ficken, J.H., Pankow, J.F., and M.E. Rosen, Effects of sample isolation and handling on the recovery of purgable organic compounds, *GWMR,* Spring, 1994.
18. Gilham, R.W., M.J.L. Robin, J.F. Barker, and J.A. Cherry, *Groundwater Monitoring and Sample Bias,* American Petroleum Institute, Washington, D.C., 1983.
19. Herzog, B. l., F.J. Chou, J.R. Valkenburg, and R.A. Grifin, Changes in volatile organic chemical concentrations after purging slowly recovering wells, *GWMR,* 8(4), 93, 1988.
20. Herzog, B., J. Pennino, and G. Nielsen, *Practical Handbook of Ground-Water Monitoring,* David M. Nielsen, Ed., Lewis Publishers, Chelsea, MI, 1991.
21. Hix, G.L., What you don't find isn't there, or is it? (or the absence of proof is not the proof of absence), *GWMR,* Winter, 1994.
22. Holm, T.R., G.K. George, and M.J. Barcelona, Oxygen transfer through flexible tubing and its effects on ground-water sampling results, *GWMR,* 8, 83–89, 1988.
23. Iles, D.L., P.D. Hammond, and L.D. Schultz, Effects of Sampling Methods on Inorganic Water Chemistry Results, National Ground Water Sampling Symposium Proceedings, Washington, D.C., November 30, 1992.

24. Imbrigiotta, T.E., J. Gibs, T.V. Fusillo, G.R. Kish, and J.J. Hochreiter, Field evaluation of seven sampling devices for purgeable organic compounds in ground water, *Ground-Water Contamination: Field Methods,* ASTM Spec. Tech. Publ. 963, American Society for Testing and Materials, Philadelphia, PA, 1988.
25. Karp, K.E., A diffusive sampler for passive monitoring of volatile organic compounds in ground water, *Ground Water,* 31(5), 735–739, 1993.
26. Kearl, P.M., N.E. Korte, and T.A. Cronk, Suggested modifications to ground water sampling procedures based on observations from the colloidal borescope, *GWMR,* Spring, 1992.
27. Koopman, F.C., Downhole Pumps for Water Sampling in Small Diameter Wells, USGS Open File Report #79-1264, U.S. Department of the Interior, Washington, D.C., 1979.
28. Knobel, L.L, and L.J. Mann, Sampling for purgeable organic compounds using positive displacement piston and centrifugal submersible pumps: A comparative study, *GWMR,* Spring, 1993.
29. Lee, G.F. and R.A. Jones, Guidelines for Sampling Ground Water, *J. Water Pollution Control Fed.,* 55(1), 92, 1983.
30. Lindorf, D.E., J. Feld, and J. Connelly, Groundwater Sampling Procedures Guidelines, Wisconsin Department of Natural Resources, Madison, 1987.
31. Loux, N.T., A.W. Garrison, and C.R. Chafin, Acquisition and analysis of groundwater/aquifer samples: Current technology and the trade off between quality assurance and practical considerations, *Int. J. Environ. Anal. Chem.,* 38, 231, 1990.
32. Martin-Hayden, J.M., G.A. Robbins, and R.D. Bristol, Mass balance evaluation of monitoring well purging. *J. Contaminant Hydrology,* 8, 225, 1991.
33. McAlary, T.A. and J.F. Barker, Volatilization losses of organics during ground water sampling from low permeability materials, *GWMR,* 7(4), 63, 1987.
34. Muska, C.F., W.P Colven, V.D. Jones, J.T. Scogin, B.B. Looney, and V. Price, Jr., *Field Evaluation of Ground Water Sampling Devices for Volatile Organic Compounds,* Proceedings of the Sixth National Symposium and Exposition on Aquifer Restoration and Ground Water Monitoring, National Water Well Association, Worthington, Ohio, 1986, 235.
35. National Council of the Paper Industry for Air and Stream Improvement, *A Guide to Ground Water Sampling,* NCASI, Tech. Bull. 362:1–123, 1982.
36. O'Melia, C.R., Kinetics of colloidal chemical processes in aquatic systems, in *Aquatic Chemical Kinetics,* W. Stumm, Ed., John Wiley & Sons, New York, 1990.
37. Panko, A.W. and P. Barth., Chemical stability prior to ground-water sampling: a review of current well purging methods, *Ground-Water Contamination: Field Methods,* ASTM Spec. Tech. Publ. 963, American Society for Testing and Materials, Philadelphia, PA, 1988.
38. Parker, T.K., Eisen, M.F., Kopania, A.A., Robertson, A., Eades, R.L., and Thomas, G., Selection, Design, Installation, and Evaluation of Dedicated Groundwater Sampling Systems: A Case Study, National Ground Water Sampling Symposium Proceedings, Washington, D.C., November 30, 1992.
39. Patterson, B.M., T.R. Power, and C. Barber, Comparison of two integrated methods for the collection and analysis of volatile organic compounds in ground water, *GWMR,* Summer, 1993.

40. Paul, C.J. and R.W. Puls, Comparison of Ground-Water Sampling Devices Based on Equilibration of Water Quality Indicator Parameters, National Ground Water Sampling Symposium Proceedings, Washington, D.C., November 30, 1992.
41. Powel, R.M. and R.W. Puls, Passive sampling of groundwater monitoring wells without purging: multilevel well chemistry and tracer disappearance, *J. Contaminant Hydrology*, 12, 51, 1993.
42. Puls, R., Improved Metals Sampling Techniques for Groundwater, USEPA Tech. Trends, EPA/540/M-91/005 No. 7, U.S. Environmental Protection Agency, Washington, D.C., 1991.
43. Puls, R.W. and M.J. Barcelona, Filtration of ground water samples for metals analysis, *Hazardous Waste & Hazardous Materials*, 6(4), 385–393, 1989.
44. Puls, R.W., R.M. Powel, D.A. Clark, and C.J. Paul, Facilitated Transport of Inorganic Contaminants in Ground Water: Part II. Colloidal Transport, EPA/600/M-91/040, U.S. Environmental Protection Agency, Washington, D.C., 1991.
45. Puls, R.W., D.A. Clark, B. Bledsoe, R.M. Powel, and C.J. Paul, Metals in ground water: sampling artifacts an reproducibility, *Hazardous Waste & Hazardous Materials*, 9(2), 149–161, 1992.
46. Puls, R.W. and R.M. Powell., Acquisition of representative ground water quality samples for metals, *GWMR*, Summer, 1992.
47. Robbins, G.A. and J.M. Martin-Hayden, Mass balance evaluation of monitoring well purging. Part I, Theoretical models and implications for representative sampling, *J. Contaminant Hydrology*, 8, 203, 1991.
48. Robin, M.J.L. and R.W. Gillham, Field evaluation of well purging procedures, *GWMR*, Fall, 1987.
49. Rosen, M.E., J.F. Pankow, J. Gibs, and T.E. Imbrigiotta, Comparison of downhole and surface sampling for the determination of volatile organic compounds (VOCs) in ground water, *GWMR*, Winter, 1992.
50. Schuller, R.M., J.P. Gibb, and R.A. Griffin, Recommended sampling procedures for monitoring well, *GWMR*, 1(2), 42, 1981.
51. Scalf, M.R., J.F. McNabb, W.J. Dunlap, R.C. Cosby, and J. Fryberger, *Manual of Ground-Water Sampling Procedures*, NWWA/EPA Series, National Water Well Association, Dublin, OH, 1981.
52. Snow, D.D., M.E. Burbach, and R.F. Spalding, Ground Water Sampler Comparison Pesticide and Nitrate Concentrations, National Ground Water Sampling Symposium Proceedings, Washington, D.C., November 30, 1992.
53. Stolzenburg, T.R. and D.G. Nichols, *Effects of Filtration Method and Sampling Devices on Inorganic Chemistry and Sampled Well Water*, Proceedings of the Sixth National Symposium and Exposition on Aquifer Restoration and Ground Water Monitoring, May 19-22, 1986, National Water Well Association, Dublin, OH, 1986.
54. Stumm, W. and J.J. Morgan, *Aquatic Chemistry*, Wiley-Interscience, New York, 1981.
55. Sung, W. and J.J. Morgan, Kinetics and product of ferrous iron oxygenation in aqueous systems, *Environ. Sci. ReTechnolo.*, 14(5), 561, 1980.
56. Trela, J.J., *Should Ground Water Samples From Monitoring Wells Be Filtered Before Laboratory Analysis? Yes and No*, Proceedings of the Sixth National Symposium and Exposition on Aquifer Restoration and Ground Water Monitoring, National Water Well Association, Dublin, OH, 1986.

57. Unwin, J.P. and D. Huis, *A Laboratory Investigation of the Purging Behavior of Small-Diameter Monitoring Wells,* Proceedings of the Third International Symposium on Ground Water Monitoring and Aquifer Restoration, National Water Well Association, Dublin, OH, 1983.
58. Unwin, J. and V. Maltby, Investigation of techniques for purging groundwater monitoring wells and sampling ground water for volatile organic compounds, *Ground-Water Contamination: Field Methods,* ASTM Spec. Tech. Publ. 963, American Society for Testing and Materials, Philadelphia, PA, 1988.
59. United States Federal Register, Oct. 9, 1991. Section 258.53, Paragraph B of the landfill rule, p. 50978.
60. USEPA, RCRA Ground-Water Monitoring Technical Enforcement Guidance Document, U.S. Environmental Protection Agency, Washington, D.C., 1986.
61. USEPA, Final Draft of Chapter Eleven of SW-846, Ground Water-Monitoring, U.S. Environmental Protection Agency, Washington, D.C., 1991.
62. Yao, K.M., M.T. Habibian, and C.R. O'Melia, Water and wastewater filtration: concepts and applications, *Environ. Sci. Technol.*, 11, 1105, 1971.
63. APHA, *Standard Methods for the Examination of Water and Waste Water,* 18th ed., American Public Health Association, Washington, D.C., 1992.
64. Puls, R.W. and R.M. Powell, Transport of inorganic colloids through natural aquifer material: implications for contaminant transport, *Envir. Sci. Technol.*, 26, 614–621, 1992.
65. Schoenleber, J.R. and P.S. Morton, Field Sampling Procedures Manual, New Jersey Department of Environmental Protection and Energy, May 1992.

chapter six

Ground water sampling plans

Jesus answered, "Everyone who drinks this water will be thirsty again, but whoever drinks the water I give him will never thirst. Indeed, the water I give him will become in him a spring of water welling up to eternal life."
The Bible, John 4:13–14 (NIV)

Introduction

To provide representative data, ground water sampling plans must be prepared by individuals who have a thorough understanding of the techniques and the operating principles of ground water sampling equipment. Ground water sampling personnel must also be adequately trained and familiar with operating the equipment specified in the sampling plan. The value of the data generated by a monitoring program is directly proportional to the individual operator's skill and familiarity with the operation of the device used for sampling.[3]

Site-specific sampling plans should clearly and systematically define all the steps required for taking ground water samples. The more specific the sampling plan, the less chance for introducing errors or erroneous assumptions.[4] Several checklists need to be prepared and used which detail the equipment maintenance required before going into the field and specify the documentation, equipment, and supplies required for sampling.

Some elements of sampling plans are common to most monitoring situations, and some are unique to the site. Sampling plans need to be "stand alone" documents taken out into the field and written so they can be understood clearly by sampling personnel with varying educational backgrounds. The following is intended to provide a broad overview of sampling plans from which specific plans may be prepared and executed.

Health and safety plans

This section does not purport to address all or even most of the elements in a site-specific safety plan. It is the responsibility of the individuals preparing the site-specific safety plan to address all the necessary elements and obtain adequate approval from appropriate regulatory agencies. As per *29 CFR 1910.120* safety and health plans are required for:

- All CERCLA sites, state-ordered cleanup operations, and initial site investigations
- All RCRA cleanup operations
- All responsible party voluntary cleanup operations at regulated uncontrolled hazardous waste sites
- All operations involving hazardous wastes conducted at treatment, storage, and disposal (TSD) facilities
- Emergency response operations

Requirements for a site safety and health plan as presented in *29 CFR 1910.120* include a risk analysis for each site task, employee training assignments, personal protective equipment to be used for each task, and medical surveillance requirements. In addition, the plan must specify: frequency and types of air monitoring, personnel monitoring, environmental sampling, site control measures, decontamination procedures, an emergency response plan, confined space entry procedures, and a spill containment plan. The Occupational Safety and Health Administration (OSHA) requires special training (40 hours of initial training and 8 hours of annual refresher training) for sampling at hazardous waste sites.

A site-specific health and safety plan must always be considered for sampling activities. Workers should not be exposed needlessly to hazardous chemicals and should be protected from accidents. A safety and health plan should include properly addressing potential accidents, heat or cold stress, biological hazards (snakes, spiders, ticks, etc.), and potential exposure to hazardous solids, liquids, and vapors. The site-specific safety and health plan may need to be reviewed by local OSHA representatives.

Protective garments and gloves, respirators or SCBA equipment, safety boots, safety glasses, and hard hats may be required for sampling at some hazardous waste sites. A generic health and safety plan that would include site-specific emergency phone numbers, maps to hospitals, types of protective garments such as gloves, boots, etc., should be generated for sampling even in relatively benign situations. If special safety and health precautions are not required, then that should be stated in the sampling protocol. The safety and health plan should be reviewed prior to going into the field and taken along to the site with the other documents.

In accordance with the site-specific safety and health plan, it is probably a good convention for sampling personnel to assume that an explosive hazard exists at any well at the site. The following precautions should be taken until proven otherwise:

1. Vehicles should be parked downwind (to avoid potential sample contamination concerns) at a minimum of 15 ft from the well and the engine turned off prior to vapor analysis.
2. Smoking should not be allowed within 15 ft of the well

3. No heat-producing or electrical instruments should be within 15 ft of the well, unless they are intrinsically safe, prior to vapor analysis
4. An explosivity and organic vapor monitor should be used to monitor gases in the well

If explosive levels of gases are present between the inner and outer casing, then adequate protection including a heat-resistant face shield and heat-resistant gloves may be warranted. If the inner well cap is not vented, a great deal of gas pressure can build up in the casing from rising water levels. When opening the inner well cap the sampling team should keep their faces away from the well. If explosive or toxic gases are suspected under these circumstances, remote opening of the metal well cap should be considered and the use of intrinsically safe equipment for subsequent steps should be required. Proper calibration and operation of explosivity and OVM meters can be difficult, and the operator must be thoroughly familiar with their operation and limitations before going into the field.

Sampling plans and checklists

Sampling plans need to be very specific about the equipment and techniques used for acquiring ground water samples in individual wells at a site. The sampling order of the wells, the exact type of equipment that will be used to purge and sample individual monitoring wells, the exact purge and sample rates to be used, the equipment and techniques proposed for field filtering, detailed field cleaning procedures, etc., must be presented in a way that avoids any ambiguity. Generic "cookie cutter" plans that are reused in their entirety for multiple sites are not acceptable. The following information should be included in a formal sampling plan:

1. The name of the site, the date the plan was prepared, and the individual(s) who prepared it
2. A description of previous site activities and potential contaminants
3. The purpose and goals of ground water monitoring at the site
4. The names and phone numbers of individuals who are responsible for monitoring activities at the site and who can approve of field changes
5. An overview of the geology and hydrogeology of the site including the ground water flow direction
6. A site map showing the locations of any potentially concentrated contaminants, the monitoring points, property boundaries, etc.
7. Table(s) including well field numbers, relative location (upgradient or downgradient), parameters to be sampled, top of riser elevations, and distances from measuring point to the top of the screen and to the bottom of the well
8. A listing of the sequence of sampling, from the least to the most contaminated well
9. A detailed description of the specific sampling equipment proposed to be used for each well at the site

10. A detailed, step-by-step description of the processes required for obtaining field parameters and ground water samples for each well at the site
11. A Quality Assurance/Quality Control component (see Chapter 7)
12. A description of how the samples are to be transported to the analytical laboratory
13. A chain-of-custody plan (if required)
14. A site-specific health and safety plan
15. A list of analytes and a schedule of when the site is to be sampled

The objective of sampling is to collect a portion of material, small enough in volume to be transported conveniently and handled in the laboratory, that accurately represents the material being sampled.[1] The following is a typical sequence for obtaining ground water samples from monitoring wells.

Before going into the field

1. Review the most current site-specific sampling plan and the health and safety plan
2. Provide adequate notification to the laboratory, site owner, and regulatory agencies that a sampling event is going to occur
3. Check and calibrate the equipment
4. Assemble documents, equipment, and supplies
5. Obtain the keys to the site and monitoring wells

Upon arrival at the first well in the sampling network

1. Inspect the well for damage and for evidence of tampering (a secure lock in place)
2. Use an explosivity meter and/or organic vapor monitor to check the outer and inner casing, as necessary
3. Prepare the area around the well for sampling
4. Calibrate the equipment in the field, as required
5. Take static water levels (and NAPL or DNAPL levels, if appropriate)
6. Measure field parameters while simultaneously purging the well
7. Adequately label the sample containers
8. Document the field work including initial, stabilizing, and final field parameters
9. Acquire the sample and field filter (if appropriate); add preservatives
10. Place the samples in a shipment cooler with ice or freezer packs
11. Prepare quality control samples
12. Perform field analyses such as for alkalinity
13. Complete the documentation for the well
14. Clean the equipment and go to the next well

15. Institute chain-of-custody controls as required
16. Ship the samples to the laboratory for analysis

Numerous permutations are possible with the above generic sequence, but many if not most sampling efforts follow in the same general order the steps listed above. Each step of the process must be adequately documented by field personnel to ensure that samples are taken properly and are legally defensible if challenged in a court of law.

The types of equipment required for ground water sampling include water level indicators, field geochemical instrumentation (pH, temperature, specific conductance, etc.), sample acquisition devices (e.g., pumps and bailers), and ancillary equipment and supplies such as deionized water, ice, hand tools, disposable gloves, etc. When potential human health and environmental impacts, analytical costs, and the costs of remedial actions (which are based on ground water samples) are considered, the sampling equipment and supplies should be of the best available quality.

Prior to going into the field

Checklists should be developed and used to so that all of the required steps prior to going into the field are undertaken. Good check lists will include:

- Providing adequate notification to the laboratory (so that holding times are not exceeded) and to the owner of the site and the primary regulatory agency (in writing) that a round of sampling is to commence in order to facilitate sampling and allow for a sampling audit or split sampling
- Specifying and documenting the equipment maintenance and calibration undertaken prior to going into the field relative to the sampling event
- Listing the documents, equipment, and supplies required to sample the site

Prior to going into the field, sampling personnel should reacquaint themselves with the sampling plan. The review is undertaken so that the required specific protocol such as sampling from the least to the most contaminated wells, knowing where quality control samples are to be taken, knowing the disposition of purge water, etc., is understood and followed.

The amount of equipment maintenance and calibration required prior to going into the field should be clearly specified in the presampling equipment maintenance and calibration checklists, which are based on the manufacturer's recommendations, sampling objectives, and prior experience. Maintenance and calibration performed before sampling must be documented to provide evidence that the equipment was adequately maintained and calibrated and to keep a permanent record of equipment servicing and performance.

A list of all the documents, equipment, and supplies required for the sampling event should be prepared and used. It can be frustrating and time consuming to forget equipment and supplies, so some up-front preparation is warranted. The following is a list of the documentation, equipment, and supplies which should assist in preparing a site-specific equipment and supply checklist.

Documentation

- Site-specific sampling plan
- Health and safety plan
- Appropriate maps
- Copies of the equipment operation manuals
- Field notebook, with sewn binding
- Stabilization forms
- Chain-of-custody forms
- Laboratory sample analysis request forms
- Shipping labels and custody seals
- Reference chart showing volume of water per unit size of casing

Equipment

- Electric water level/interface indicator and a steel tape and chalk or indicator paste(s)
- Pumps, sample tubing lines, flow controllers, power cord(s), batteries, compressors, etc.
- Bailers, rope
- Flow-through cell
- pH-Eh meter and probes, conductivity meter, dissolved oxygen meter, turbidity meter(s), calibration solutions, batteries, etc.
- Garden hose (for domestic well sampling)
- Explosivity and/or organic vapor monitor and calibration supplies
- Complete set of hand tools including a sharp knife, screw drivers, pliers, hacksaw, flashlight, large pipe wrench, hammer, bolt cutters, and replacement locks
- Field filtering equipment and supplies
- Maximum-minimum thermometers for sample shipment
- Large plastic boxes for the equipment and supplies
- Several types of scrub brushes
- Squirt bottle(s)
- Garden sprayer or power cleaner for field cleaning
- Appropriate health and safety equipment
- 5-Gallon bucket
- Large plastic garbage can for field cleaning of pumps
- Sample bottles (with extras) and preservatives
- Shipping containers (coolers)
- Portable field table and chair (optional)

- Clipboard
- Calculator
- Water resistant clock or watch with a second hand
- First aid kit

Supplies

- Laboratory grade nonphosphate detergent
- Disposable gloves, appropriate to the contaminants
- Bags of ice or frozen freezer packs
- Plastic garbage bags
- Plastic sheeting
- Sufficient quantities of potable and laboratory grade deionized water for cleaning and equipment blanks
- Methanol
- Clean rags and paper towels
- Electrical tape, duct tape, and wide transparent tape
- Hand soap
- Regular, ballpoint, and indelible pens

After providing adequate notification, performing the presampling maintenance and calibration, obtaining the site and well keys, and packing the supplies and equipment, the monitoring well network is ready to be sampled.

Upon arrival at the first well in the sampling network

As appropriate, an approved safety and health plan must be prepared and used for all laboratory and field activities. Sampling personnel should not use perfume, insect repellent, hand lotion, etc., when taking ground water samples. If insect repellent must be used, then sampling personnel should not allow samples or sampling equipment to contact the repellent, and it should be noted in the documentation that insect repellent was used. Smoking and eating should not be allowed until the well is sampled and hands are washed with soap and water, due to safety and possibly sample contamination concerns. The order of sampling should be from the least contaminated to the most contaminated well to reduce the potential for cross contamination of sampling equipment. The order of sampling needs to be specified in the sampling plan.

Inspect the well for damage

The first thing required when arriving at the site is to inspect the well for damage. Wells that show cracked aprons or bent casings, are unlocked, or show other signs of damage or tampering may not warrant any further sampling effort. If damaged or unsecured wells are encountered, they should be immediately reported, as the well may need to be replaced prior to sampling.

Use an explosivity meter or organic vapor monitor to check the head space in the well

At some sites it may be necessary to use an organic vapor monitor (OVM) and/or an explosivity meter before undertaking other sampling work at the site. The OVM and explosivity meter are used for a first indication of the presence of light nonaqueous phase liquids (LNAPLs), and to monitor the potential for fire, explosion, or toxic effects on workers.[7]

Prepare the area around the well for sampling

The next step in sampling a well is to prepare the immediate work area by removing any proximal obstructions (including weeds) and laying down a new piece of plastic sheeting on the ground to keep equipment from being contaminated by coming into contact with the ground surface. A portable field table covered with a new piece of plastic for each well is convenient to use for preparing equipment and performing field measurements. Sampling equipment must not be placed on the ground, because the ground may be contaminated and soil contains trace metals. Equipment and supplies should be removed from the vehicle only when needed.

Calibrate equipment in the field, as required

The sampling plan must specify equipment that needs to be calibrated in the field and how often after the initial calibration it should be checked or recalibrated. Some of the instrumentation used for obtaining field parameters should remain turned on for the duration of the sampling event to obviate a warm-up period and initial recalibration.

Stock calibration solutions for pH meters are available with a pH of 4, 7, and 10. To ensure traceability of the calibration solutions, the source (lot number, date, etc.) of the calibration solutions should be documented in the field notes. A pH meter must be periodically calibrated with a two-point calibration by using two buffer solutions that bracket the expected pH of the ground water. For example if the anticipated pH of the ground water is 6.5, the calibration solutions should have pHs of 4 and 7. Fresh calibration solutions should be used for each sampling event.

Calibration of dissolved oxygen meters may be involved and should be redone in the field at least once per day and possibly more often if changes in elevation or atmospheric pressure occur. Checking and documenting the performance of a electronic dissolved oxygen meter against a titration method at least once per day is recommended.

A conductivity meter should be checked with standard solutions prior to going out in the field; if the reading is out of prescribed tolerances it may need servicing, as readjustments of the meter may not be possible. Checking and documenting the performance of conductivity meters may be done in the field with two audit solutions.

Take static water levels (and LNAPL or DNAPL levels if required)

Recording water levels in all the wells at the site prior to taking any samples may be required, especially at larger sites or at sites where water levels may change relatively rapidly. All the static water levels must be taken on the same day and not held over from one day to the next. Determining the thickness of the LNAPL and DNAPL are required for both the field notes and for calculating the volume of the product that needs to be removed before sampling the ground water. If the well has a history of contamination, then disposable gloves should be worn while making water level measurements. If the concentrations of contaminants are approaching that of free product, then upgrading the quality of the disposable gloves (and other clothing and equipment) may then be warranted.

If dedicated equipment was installed, it may not be possible to measure the depth to the bottom of the well, because the water level indicator may not fit around the dedicated pump. Using the measured depth to water, and knowing the length of the well screen, the volume of stagnant water above the screen can be calculated and potentially used for purging criteria as per the sampling plan. Periodically (specified in the sampling plan), the dedicated pump and lines should be pulled for inspection, servicing, and replacement. At that time the degree of siltation into the well needs to be determined and recorded using a water level indicator, and the sediment be removed as required.

If dedicated equipment was *not* installed, then the amount of water in the casing is measured and calculated for stabilization criteria of the well. The degree of siltation in the well can be readily ascertained by comparing the measured depth to the depth recorded on the well installation record and by the feel of the indicator as it drops onto the bottom cap of the well.

Obtain field parameters while simultaneously purging the well

A flow-through cell must be used to obtain reliable field parameters with properly calibrated equipment. Flow-through cells must not be pressurized or damage to the probes may occur. When the sampling pump is turned on, the initial ground water field parameters (temperature, pH, specific conductance, dissolved oxygen, turbidity) are recorded. Field parameters are then periodically recorded, and when the predetermined criteria of well stabilization as per the sampling plan are met, the well has been adequately purged and sampling may commence. A field stabilization form should be used to document the purging and to record the final field parameters (Figure 6.1). Variations of this form may be prepared for differing site objectives and may need approval from regulatory agencies.

Adequately label the required bottle types

The sampling plan should identify the types of containers to be used for sample collection. Containers used for obtaining samples must be inert with

DATE & TIME: ____________ STATION: ____________

SAMPLES COLLECTED BY: ____________

FIELD CONDITIONS, OBSERVATIONS, NOTES: ____________

WELL TYPE: ____________ WELL SIZE (INNER DIAMETER): ____________

CONDITION (LOCKED, DAMAGED ETC.): ____________

PARAMETERS SAMPLED FOR: ____________

SAMPLE APPEARANCE (COLOR, SMELL, ETC.): ____________

FREE PRODUCT THICKNESS (IF APPLIC.): ____________ OVM/EXP. READING (IF APPLIC.) ____________

DEPTH TO WATER (0.01 FT.): ____________ DEPTH TO BOTTOM (0.01 FT.): ____________

VOLUME OF WATER IN WELL BORE: ____________ CALCULATED PURGE VOLUME: ____________

ACTUAL PURGE VOLUME: ____________ METHOD OF PURGE: ____________ PURGE RATE: ____________

SAMPLING METHOD: ____________ SAMPLE RATE: ____________

SAVE PURGE WATER? (Y/N): ____________ IF SAVED CONTAINMENT METHOD: ____________

DESCRIBE FIELD CLEANING OF EQPT.: ____________

STABILIZATION TRIALS: 3 CONSECUTIVE READINGS WITHIN SPECIFIED LIMITS)

	TIME:	TEMP.:	pH:	COND.:	Eh:	D.O.:	TURB.:
1.							
2.							
3.							
4.							
5.							
6.							

FINAL VALUES TEMP.: ________ pH: ________ COND.: ________ Eh: ________ D.O.: ________ TURB.: ________

ALKALINITY: ____________ OTHERS (SPECIFY): ____________

FILTERED (Y/N): ____________ IF YES SPECIFY METHOD: ____________

PRESERVATIVES (Y/N): ____________ IF YES SPECIFY: ____________

QA/QC SAMPLE TAKEN (Y/N)?: ____________ IF YES SPECIFY: ____________

CHAIN OF CUSTODY DONE? (Y/N): ____________ IF YES CHAIN OF CUSTODY NUMBER: ____________

SAMPLE NUMBER: ____________ NAME OF ANALYTICAL LABRATORY: ____________

Figure 6.1 A typical field stabilization form.

respect to the parameters to be sampled. They also should be properly cleaned. This is important due to the length of time the sample will be in contact with the walls of the container.

The preservative must also be accounted for in the bottle specifications. Ground water samples to be analyzed for trace metals are preserved by addition of nitric acid to a pH of below 2. It has been reported that up to 350 µg/L of lead has leached from some sample bottles after 19 days, due to adjusting the pH to below 2, through the use of incorrect bottles.[8] The bottles recommended for sample collection are provided in Table 6.1.

Analytical laboratories may not accept samples for analysis if the bottles have not been cleaned by their own laboratory for QA/QC concerns. The following is an example of the procedure used for cleaning VOC vials:

Bottles

1. Remove the label and pour out any remaining sample water
2. Separate the cap and septum, rinse them with tap water, and soak them in a dishpan of warm, soapy water, using laboratory grade nonphosphate detergent
3. Scrub the vials with a brush
4. Rinse the vials with tap water until the soap residue is gone, followed by a triple rinse of deionized water
5. Place the vials in a metal basket right-side-up and place in a ventilated oven at 105°C for at least 3 hours, but optimally overnight
6. Remove the vials from the oven and allow at least 20 min to cool
7. Each vial is then capped with a heated septum and stored until needed

Caps

1. Wash the caps in soapy water to remove surface dirt
2. Rinse with tap water to remove soap, allow to air dry, and store until needed

Septum

1. Wash in soapy water, separate from the other pieces
2. Rinse with tap water to remove soap residue and then with a triple rinse of deionized water
3. Spread the septum in a single layer on a cleaned stainless steel tray lined with laboratory grade tissue paper
4. Dry in an oven at 105°C for no more than an hour
5. Remove, cool, and use immediately to cap vials

Sample containers for semivolatile organic compounds (pesticides, herbicides, etc.) should be soap-and-water washed, followed by a methanol or isopropanol rinse.

Assess the adequacy of sample container cleaning by filling a certain number of the cleaned bottles with laboratory grade deionized water and then analyze the performance sample as part of the laboratory's QA/QC plan. While the well is being purged, the bottles should be adequately labeled. Labels on sample bottles should not come off and should remain legible even when wet. When sampling for VOCs, the pen's ink may cause false positives, so the labels should be filled out and the ink allowed to dry before being affixed to the bottles. The minimum amount of information on bottle labels includes:

Table 6.1 Containers, Preservation Techniques, and Holding Times for Aqueous Matrices

Name	Container[a]	Preservation	Maximum holding time
Bacterial Tests:			
Coliform, total	P, G	Cool, 4°C, 0.008% $Na_2S_2O_3$	6 hours
Inorganic Tests:			
Chloride	P, G	None required	28 days
Cyanide, total and amendable to chlorination	P, G	Cool, 4°C; if oxidizing agents present add 5 mL 0.1N $NaAsO_2$ per L or 0.06 g of ascorbic acid per L; adjust pH>12 with 50% NaOH. See Method 9010 for other interferences.	14 days
Hydrogen ion (pH)	P, G	None required	24 hours
Nitrate	P, G	Cool, 4°C	48 hours
Sulfate	P, G	Cool, 4°C	28 days
Sulfide	P, G	Cool, 4°C, add zinc acetate	7 days
Metals:			
Chromium VI	P, G	Cool, 4°C	24 hours
Mercury	P, G	HNO_3 to pH<2	28 days
Metals, except chromium VI and mercury	P, G	HNO_3 to pH<2	6 months
Organic Tests:			
Acrolein and acrylonitrile	G, Teflon(TM)-lined septum	Cool, 4°C, 0.008% $Na_2S_2O_3$[c], adjust pH to 4-5	14 days
Benzidines	G, Teflon-lined cap	Cool, 4°C, 0.008% $Na_2S_2O_3$[c], adjust pH to 6-9, store in dark	7 days until extraction, 40 days after extraction
Chlorinated hydrocarbons	G, Teflon-lined cap	Cool, 4°C, 0.008% $Na_2S_2O_3$[c]	7 days until extraction, 40 days after extraction

Dioxins and Furans	G, Teflon-lined cap	Cool, 4°C, 0.008% $Na_2S_2O_3$[c]	7 days until extraction, 40 days after extraction
Haloethers	G, Teflon-lined cap	Cool, 4°C, 0.008% $Na_2S_2O_3$[c]	7 days until extraction, 40 days after extraction
Nitroaromatics and cyclic ketones	G, Teflon-lined cap	Cool, 4°C, 0.008% $Na_2S_2O_3$[c] store in dark	7 days until extraction, 40 days after extraction
Nitrosamines	G, Teflon-lined cap	Cool, 4°C, 0.008% $Na_2S_2O_3$[c] store in dark	7 days until extraction, 40 days after extraction
Oil and grease	G	Cool, 4°C[b]	28 days
Organic carbon, total (TOC)	P, G	Cool, 4°C[b]	28 days
PCBs	G, Teflon-lined cap	Cool, 4°C	7 days until extraction, 40 days after extraction
Pesticides	G, Teflon-lined cap	Cool, 4°C	7 days until extraction, 40 days after extraction
Phenols	G, Teflon-lined cap	Cool, 4°C, 0.008% $Na_2S_2O_3$[c]	7 days until extraction, 40 days after extraction
Phthalate esters	G, Teflon-lined cap	Cool, 4°C	7 days until extraction, 40 days after extraction
Polynuclear aromatic hydrocarbons	G, Teflon-lined cap	Cool, 4°C, 0.008% $Na_2S_2O_3$[c] store in dark	7 days until extraction, 40 days after extraction
Purgeable aromatic hydrocarbons	G, Teflon-lined septum	Cool, 4°C, 0.008% $Na_2S_2O_3$[b,c]	14 days
Purgeable Halocarbons	G, Teflon-lined septum	Cool, 4°C, 0.008% $Na_2S_2O_3$[c]	14 days
Purgeable organic halides (TOX)	G, Teflon-lined cap	Cool, 4°C[b]	28 days
Radiological Tests:			
Alpha, beta, and radium	P, G	HNO_3 to pH<2	6 months

Note: Table excerpted, in part, from Table 11, 49 FR 209, October 26, 1984, p 28.

[a] Polyethylene (P) or Glass (G)

[b] Adjust to pH<2 with H_2SO_4, HCL, or solid $NaHSO_4$.

[c] Free chlorine must be removed prior to addition of HCl by the appropriate addition of $Na_2S_2O_3$.

Figure 6.2 Glass, polyethylene, and Teflon™ bottles used for collecting ground water samples.

- The sampler's organization (i.e., name of private sampling company or state or federal agency)
- The name of the site
- The sample point (monitoring well number, residential well, etc.)
- Date and time
- Sampler's initials

Providing other information such as the sample preservative, the number of bottles (1 of 4, 2 of 4, etc.), and the analytical parameters may be facilitated by using predesigned computerized (bar-coded) labels, or, if there is space, this information may also be written on the bottle labels with an indelible pen (Figure 6.2).

Document the field work including initial, stabilizing, and final field parameters

The three words used to ensure adequate documentation for ground water sampling are accountability, controllability, and traceability.[6] Accountability is undertaken in the sampling plan and answers the questions who, what, where, when, and why to assure that the sampling effort meets its goals. Controllability refers to checks (including QA/QC) used to ensure that the procedures used are those specified in the sampling plan. Traceability is documentation of what was done, when it was done, how it was done, and by whom it was done, and is found in the field forms, field notebook, and chain-of-custody forms. At minimum, adequate documentation of the sampling that was done in the field consists of an entry in a notebook with a sewn binding, a field parameter-stabilization sheet for each well, a sample analysis request sheet, and possibly a chain-of-custody form.

As a general rule, if one is not sure whether the information is necessary, it should nevertheless be recorded, as it is impossible to over-document one's field work.[3] Years may go by before the documentation comes under close scrutiny, so the documentation must be capable of defending the sampling effort without the assistance or translation of the sampling crew.

The minimum information to be recorded daily with an indelible pen in a notebook with a sewn binding includes date and time(s), name of the facility, name(s) of the sampling crew, site conditions, the wells sampled, a description of how the sample shipment was handled, and a QA/QC summary. After the last entry for the day, the crew chief should sign the bottom of the page under the last entry and then draw a line across the page directly under the signature.

Computerized field data input with a durable, water- and chemical-resistant hand-held or lap-top computer may facilitate data entry and downloading. Ensuring that the data are backed up and that provisions have been made against data tampering need to be considered when using a computerized system. The use of computerized data entry in conjunction with bar-coded sample bottles can be effective to track a sample from collection through analysis, and data reporting.

Electronic or paper stabilization forms are used to record who did the sampling, to record observations on the site conditions, to assist in documenting nonconservative physical and chemical information, and to ensure that the specified protocol was followed. The sampling crew must provide copies of the original, unedited field notes to the analytic laboratory for inclusion with the analytic results. Copies of the original field notes are needed to provide information on static water levels and stabilized field parameters and to ensure conformance with the specified protocol.

Acquire the sample, field filter (if appropriate), add preservatives

The sampling plan must specify the order in which samples are taken. The general order of sampling if no free product is present[7] is

1. Field parameters
2. Volatile organics (VOA)
3. Purgeable organic carbons (POC)
4. Purgeable organic halogens (POH)
5. Total organic halogens (TOX)
6. Total organic carbon (TOC)
7. Base neutrals/acid extractable
8. Total petroleum hydrocarbons and oil and grease
9. PCBs and pesticides
10. Total metals

Figure 6.3 A 40-mL purge-and-trap vial with a positive meniscus (right) and a 40-mL purge-and-trap vial without a positive meniscus (left).

11. Phenols
12. Cyanide
13. Sulfate and chloride
14. Turbidity
15. Nitrate-N and ammonia
16. Preserved inorganics
17. Radionuclides
18. Unpreserved inorganics
19. Bacteria

This collection order takes into account the potential volatilization of components within the ground water sample.

When volatile organic compounds (VOCs) are sampled, the bottles should not be rinsed with sample water prior to filling. Reagent grade preservative supplied by the analytical laboratory is added to the 40-mL purge and trap bottles, and the bottles are filled so as to leave a positive meniscus (Figure 6.3).

The Teflon™ septum caps are then screwed on quickly over the positive meniscus to preclude any air bubbles from entering the bottle. Each bottle is then turned over and tapped (flicked with the finger) to see if any air bubbles were trapped in the sample. If air bubbles are detected, the sample should be discarded and a new sample obtained (Figure 6.4). Extra labels, vials, septa, and caps must be provided to the sampling crew by the analytical laboratory

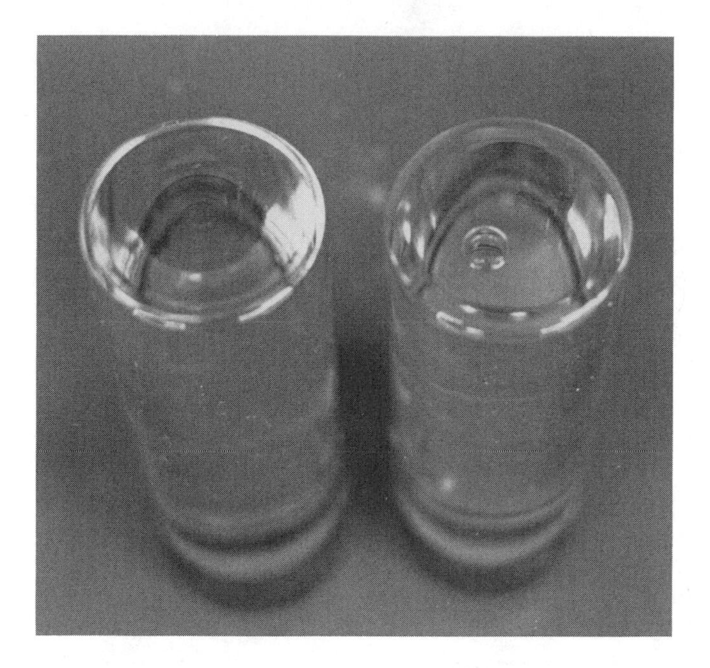

Figure 6.4 Two upside-down 40-mL purge-and-trap vials; the one on the right has an unacceptable air bubble.

to replace those invalidated by the presence of air bubbles, a loose-fitting septum, or other errors.

Ambient odors, vehicle exhaust, precipitation, or windy/dusty conditions can potentially interfere with obtaining representative samples. These conditions should be minimized and should be recorded in the field notes. The outlet from the sampling device should discharge below the top of the sample's air/water interface, when possible. The sampling plan should specify how the samples will be transferred from the sample collection device to the sample container to minimize sample alterations.[7] Samples should not be transferred from one sampling container to another.[8]

The amount of time that sample containers remain open needs to be minimized, and sampling personnel must not touch the inside of sample bottles. Sample bottles must be shielded from strong winds, rain, and dust when being filled.

When taking ground water samples (especially for VOCs and trace metals), sampling personnel should use a new pair of disposable gloves for each well sampled to reduce the potential for exposure of the sampling personnel to contaminants and to reduce sample contamination. It may be a good practice to change disposable gloves between purging and sampling operations at the same well, depending on the specifics of the site.

From a safety perspective, using gloves with holes in them may be worse than not using gloves at all. Disposable gloves worn by sampling personnel are usually made of latex or vinyl. Some disposable gloves have a fine layer of corn starch on the inside of the gloves to facilitate putting them on and taking them off. Disposable gloves may be a source of phthalates, which could be introduced into ground water samples if the gloves contact the sample. In some sampling situations, the use of more chemically resistant gloves may be warranted.

As soon as possible after collection, ground water samples need to be stabilized (preserved) to reduce chemical and biological activity from altering the sample. Sample preservation is intended to retard biological action and hydrolysis and to reduce sorption effects.[7] Preservation techniques can reduce, but cannot completely stop, chemical and biological processes. Sample preservation includes temperature and pH controls, chemical addition, and protection from light. Table 6.1 provides the recommended preservation techniques for various chemical constituents.

Chemical preservatives have a relatively short shelf life and should be supplied fresh by the analytical laboratory, which may need to supply the sample bottles as part of their internal QA/QC. Chemical and biological activities that alter samples[3] include one or more of the following:

- Formation of complexes
- Adsorption/desorption
- Acid-base reactions
- Oxidation-reduction reactions
- Precipitation-dissolution reactions
- Microbiological activities which affects the disposition of metals, anions, and organic molecules

Ground water samples obtained for general chemistry are not chemically preserved. Ground water samples obtained for metals are chemically stabilized by the addition of nitric acid (HNO_3) to a pH of 2, and samples obtained for nutrient analysis (nitrogen and phosphorus) are chemically stabilized with sulfuric acid (H_2SO_4). All chemical preservatives need to be supplied by the analytical laboratory to ensure their purity. The preservative bottles should be cleaned and the reagents replaced at least quarterly by the laboratory.

Whenever possible, chemical preservatives should be added to cleaned containers in the laboratory, but they may also be added to the sample in the field. Sampling containers containing preservatives added by the laboratory should not be overfilled or dilution of the preservative will occur. As per the approved site-specific safety plan, gloves and splash-proof goggles should be worn by sampling personnel when using preservatives.

In water samples with low chloride content, rapid transformations of VOCs have been observed. The addition of sodium bisulfate to VOC samples may preserve the samples over a longer time and would allow direct comparison of samples from quarter to quarter.[5] This preservation technique would need regulatory approval.

Place the samples in a shipment cooler

Temperature control (part of sample preservation) is usually undertaken by placing the filled sample containers in a hard-cased cooler (which is to be used to ship the samples) with ice and kept at 4°C until received by the laboratory. Replenishment of the ice or replacement of frozen freezer packs may be required in some situations. A maximum-minimum temperature recording thermometer should be included with the samples to ensure that the samples are kept at the proper temperature until delivery to the laboratory. The laboratory should record the maximum-minimum temperatures upon receipt of the cooler. The bottles must not be allowed to freeze or they may break. Hard-cased coolers are used to keep the samples as close to 4°C as possible, to reduce the risk of bottle breakage or loss, to keep the samples out of the light, and to reduce the potential of tampering. Shipment coolers need to be packed very carefully and adequately labeled.

Prepare quality assurance/quality control samples as appropriate

The sampling plan will dictate which type and how many quality control samples need to be taken (see Chapter 7). At minimum, a trip blank and at least one field duplicate for each sampling event should be required.

Perform field analyses such as alkalinity, turbidity, dissolved oxygen

Alkalinity is undertaken by titration (see Chapter Five), and since alkalinity can change substantially between the field and laboratory it must be done in the field. For precise measurements, triple analysis and an averaging of the results is recommended.

Turbidity is measured either continuously or in discrete aliquots in the field with a nephelometer. Turbidity has been demonstrated to correlate with well stability and with sample representativeness, so field turbidity measurements should be required for each well for each sampling event.

Dissolved oxygen measurements are either made continuously as a stabilization indicator or may be taken similarly to alkalinity at the end of the sampling event. Dissolved oxygen should be a required parameter for ground water sampling as several geochemical reactions depend upon it.

Complete the documentation for the well

A completed assemblage of paperwork for a sampling event includes the completed field stabilization forms, entries in the field notebook (with a sewn binding), laboratory analysis request sheet(s), transportation documentation (if required), and possibly chain-of-custody forms. If the sample could be hazardous to the analytical laboratory personnel or could damage the laboratory's instrumentation, then this must be clearly documented on the request-for-analysis sheet.

Clean the equipment and go to the next well

The site-specific sampling plan will specify the required field cleaning. Whenever possible, sample acquisition equipment should be cleaned in the laboratory, used once in the field, and then returned to the laboratory for cleaning. Field cleaning is described in the previous chapter. The field cleaning procedure used needs to be adequately documented in the field notes.

Chain-of-custody controls

A chain-of-custody procedure establishes the documentation and control necessary to identify and trace a sample from its collection to final analysis. Chain-of-custody is necessary if there is any possibility that analytical data or conclusions based upon analytical data will be used in litigation.[2] Chain-of-custody procedures and documentation begin at sample collection and continue through shipment to acceptance and analysis at the laboratory and finally to sample destruction. The primary objective is to create an accurate written record that can be used to trace the possession of the sample from the moment of its collection through to its introduction into evidence in a court of law.

As per the USEPA,[7] a sample is considered to be in custody if it is

- In actual physical possession
- In view, after being in physical possession
- In physical possession and locked up so that no one can tamper with it
- In a secured area, restricted to authorized personnel only

Strict chain-of-custody documentation includes:

- Adequate sample labels to prevent mix-ups
- Container seals to prevent unauthorized tampering
- Field notes/logbook pertaining to sample collection
- The chain-of-custody record
- Sample analysis request sheets
- Laboratory logbook and analysis notebook

Each time the samples are relinquished and subsequently accepted by another party who takes custody of the samples, the date, time, and signature (in ink) of the party receiving the shipping containers are required. Both the transferer and transferee should keep a signed receipt at each transfer.

Chain-of-custody forms need to be sequentially numbered so that a unique number is used for each shipment. The original chain-of-custody document should accompany the samples and a copy should be retained by the sampling crew. A bill of lading or receipt from the shipper should be retained as part of the chain-of-custody documentation. Along with the analytical results and field notes, a completed chain-of-custody form should be included.

Sample coolers must be locked in a secure place when out of physical sight of the sampling or delivery personnel (e.g., put the sample coolers in locked rooms or locked automobile trunks), or they may be invalidated in a court of law. If the sample container(s) are left in a secured area (such as a locked vehicle) the date, time, and location of the secured area should be documented on the form. Each individual in the process should be prepared to testify in a court of law that the sample was either in actual possession or adequately secured.

When samples leave the owner/operator's immediate control (such as when shipped by air freight) numbered chain-of-custody seals need to be attached to the front and back of the lid on the shipping container and possibly to individual bottles within the shipping container (Figure 6.5). Chain-of-custody seals are numbered labels with self-adhesive backs and are signed by the individual who is in charge of sampling the wells. The seals are placed across the lid of the cooler in such a manner that if the container is opened the seals are ripped. Wide, transparent sealing tape is usually wrapped around the width of the cooler and the seal to make opening the cooler by unauthorized individuals more difficult.

A sample analysis request sheet must accompany the samples to the laboratory. A typical sample analysis request sheet includes the following information:

- Name and address of the analytical laboratory
- The date the sample was collected and received by the laboratory
- Container types and numbers
- Sampling personnel name(s)
- Chain-of-custody number, if appropriate
- Description of the sample (i.e., residential well, monitoring well, etc.)
- Analyses to be performed
- Preservatives added in the field
- Special notes (field or laboratory), as required

Anything that may be important to the analyst must be clearly recorded and not just verbally conveyed, as the analyst may not be the person who accepts the samples. Incomplete documentation may result in the sample not being properly handled or damage to analytic equipment.

Sample shipment

The shipment must be delivered to the laboratory within the maximum holding times of each sample. Maximum holding times must account for

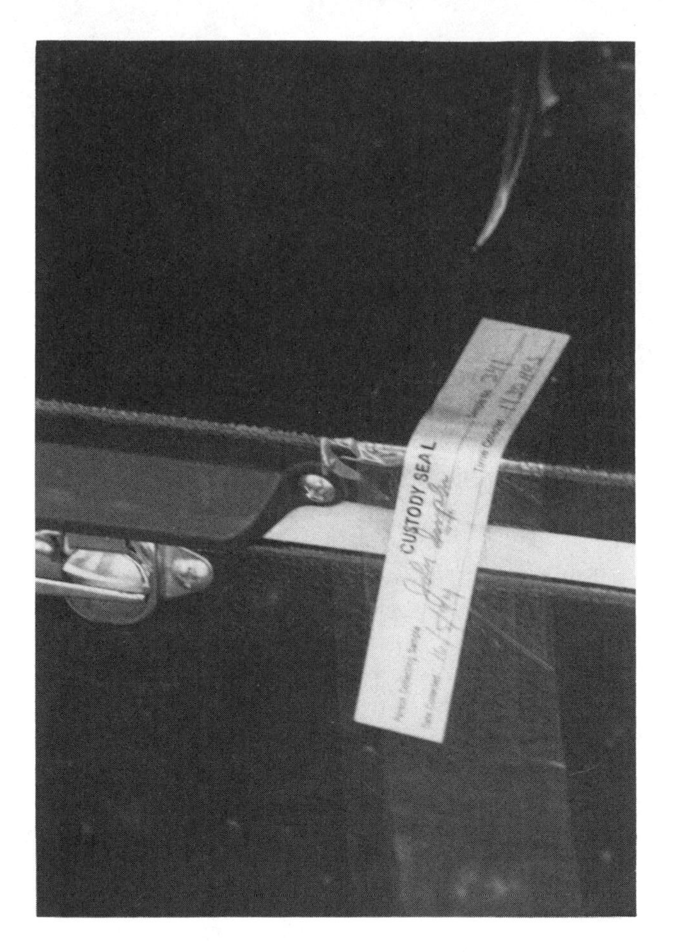

Figure 6.5 A chain-of-custody seal taped across a shipping cooler.

sample shipment and storage and analysis at the laboratory. Coordination with the laboratory must be undertaken prior to sampling to ensure that holding times are not exceeded.

When samples are transported by commercial carrier (air or surface transportation) the risks of breakage, loss, or exceeding the holding time are increased. When possible, the sampling crew should deliver the samples to the laboratory themselves and the number of people handling the samples should be kept to a minimum. If samples are considered hazardous to transport (i.e., flammable, corrosive or caustic, radioactive, etc.) special shipping procedures will be required.

U.S. Department of Transportation (DOT) regulations are legally binding and must be followed. Environmental samples must be adequately packaged so the samples do not leak or break and must have sufficient documentation. The federal rules that govern the transport of samples are found in *49 CFR*

171-179. Sample types not controlled by DOT regulations include unpreserved and preserved environmental samples that meet the following criteria:

- HCl in water solution 0.04% by weight or less
- $HgCl_2$ in water solution 0.004% or less
- HNO_3 in water solution 0.35% or less
- H_2SO_4 in water solution 0.35% or less
- NaOH in water solution 0.08% or less
- H_3PO_4 in water solution with a pH between 2 and 4

The minimum requirements for nonregulated samples (such as uncontaminated soil samples) are found in *40 CFR 261.4d2ii A* and *B.* At minimum, the sample collector's name, address, and telephone number; the laboratory's name, mailing address, and telephone number; the quantity of sample; the date of shipment; and a description of the sample must be included with the samples.

Hazardous samples are defined as radioactive materials, poisonous liquids or gases, flammable or nonflammable gases, pyrophoric material, self-reactive material, flammable liquids or solids, corrosive materials, oxidizers, combustible liquids, and miscellaneous hazardous materials *(49 CFR 173.2a).* Hazardous samples must be shipped as follows:

1. Sample jars properly labeled and sealed
2. Each sample jar placed in a plastic bag
3. Jars put into a paint can packed with vermiculite
4. Paint can put into a cooler with vermiculite and ice
5. Shipping papers and request for analysis forms placed inside the cooler
6. Custody seals placed on the cooler
7. Shipper's certification included
8. Samples shipped by the fastest available method

If chain-of-custody was instituted, upon receipt of the cooler the laboratory custodian checks the number of the intact seal on the cooler against the number(s) on the chain-of-custody form. Upon acceptance, the custodian must also record the date and time of acceptance, the internal temperature of the shipping container, and the condition of the samples in the container. In addition to logging in the sample shipping container(s), the laboratory custodian assigns laboratory numbers and stores the samples in a refrigerated, secured sample storage area until they are assigned for analysis.

All the steps including sample preparation techniques (e.g., extraction) and instrument methods must be clearly documented in the laboratory logbook. Experimental conditions, such as the use of specific reagents, temperatures, reaction times, and instrument settings should be noted. The results of the analyses of all the laboratory quality control samples should be identified, specific to each batch of ground water samples analyzed. The laboratory logbook should include the time, date, and name of the person who performed each processing step.[8]

The chain-of-custody records must be filed in a secure place for the period of time specified by applicable regulations, contractual obligations, or data use requirements, whichever is longer.

Sampling domestic and production wells

Prior to sampling domestic or high-capacity industrial or municipal wells, information about the well and distribution system should be obtained from the well owner. This information includes the location, depth of well, well construction materials, diameter of the well, and well driller. Other information, such as the drilling log and construction information, is helpful in ascertaining the hydrogeologic characteristics and limitations of the well.

Prior to sampling, an inspection (including taking adequate notes) of the well and distribution system should be undertaken and discussed with the owner. This should include observing the condition of the well (i.e., a loose-fitting well cap, graded land surface which facilitates standing water and subsequent drainage along the casing, excessive casing corrosion, etc.). A small scale map of the property with all the salient features must be entered into the notebook, including proximal sources of contamination (feed lots, septic systems, fuel tanks, etc.).

Taking static water levels (and sounding the depth to the bottom of the well) from domestic wells may have less utility than from monitoring wells or piezometers. The construction of domestic wells (especially older wells) may not be optimal for taking meaningful water levels, as the well may bridge several hydrologic units and may not be tightly cased. The measuring point for domestic wells may not be surveyed, and vertical elevations of the top of the measuring point are sometimes taken from topographic maps, which is inaccurate. Use of global positioning system equipment may facilitate horizontal and vertical controls for domestic wells.

Trying to lower a probe around a well pump may be difficult and the tape may get hung up. The spread of iron bacteria may also be facilitated if water level indicators are not adequately cleaned between wells. Some protocols call for chlorinating the well after water levels are taken to reduce the potential for introducing bacteria into the well, and allowing the chlorine to stand in the well and distribution system for a minimum of 24 hours after chlorination. Extreme care must be exercised so as not to damage or contaminate domestic wells during the sampling effort.

Samples must be collected from a cold water spigot ahead of any treatment units including water-softening equipment, whole-house filters, etc., and, if possible, ahead of the pressure tank. This is undertaken to ascertain the conditions of the water supply coming into the house and to reduce the potential for sample alterations which result from any water treatment or from the home distribution system.

In many situations a spigot may not be ahead of the pressurized storage tank. Estimating purge volumes from the well casing, plumbing, and storage tank and listening for the pump or electrical circuit to activate should be undertaken for sampling domestic wells. Thirty minutes or more may be required to empty the pressure tank.

Attaching a length of clean garden hose to a designated outside spigot (use the same spigot for subsequent sampling events) and running the purge water away from the house is recommended. Usually, outside taps are not connected to water treatment systems, but are usually downstream of the pressure tank; they are more readily accessible for sampling crews and their equipment, the home owners do not have to be home in order to take samples (prior permission is required, however), and larger volumes can be more quickly run through the system than if an inside faucet is used.

The geochemical stabilization parameters outlined in Chapter 5 are used to determine when stabilization has occurred. The geochemical stabilization parameters are recorded, the hose is disconnected, and samples are then taken from the spigot. If an inside faucet is used then the screen installed at the end of the faucet should be removed prior to sampling.

Sampling high-capacity industrial or municipal wells should be undertaken directly from the well head while the well is pumping. This reduces the potential for mixing water from different wells and changes from within the distribution and treatment system. Wells used for drinking purposes (domestic and municipal) are usually constructed in an aquifer with low total suspended solids, so samples usually do not require filtering.

References

1. APHA, *Standard Methods for the Examination of Water and Wastewater,* 18th ed., American Public Health Association, Washington, D.C., 1992.
2. ASTM, *Standard Practice for Sampling Chain of Custody Procedures,* Designation D 4840-88, American Society for Testing and Materials, Philadelphia, PA, 1988.
3. Herzog, B., Pennino, J., and Nielsen, G., *Practical Handbook of Ground-Water Monitoring,* David M. Nielsen, Ed., Lewis Publishers, Chelsea, MI, 1991, chap. 11.
4. Keith, L.H., *Environmental Sampling and Analysis: A Practical Guide,* Lewis Publishers, Chelsea, MI, 1991.
5. Maskarinec, M.P., Johnson, L.H., Holladay, S.K., Moody, R.L., Bayne, C.K., and Jenkins, R.A., Stability of volatile organic compounds in environmental samples during transport and storage, *Environ. Sci. Technol.,* 24(11), 1665, 1990.
6. Seanor, A.M., Monitoring report: assuring the quality of ground water samples, *Ground Water Age,* 18(8), 41, 1984.
7. USEPA, RCRA Ground-Water Monitoring Technical Enforcement Guidance Document, U.S. Environmental Protection Agency, Washington, D.C., 1986.
8. USEPA, Newsletter Quality Assurance, Vol. 11(2), U.S. Environmental Protection Agency, Washington, D.C., 1990.

chapter seven

Quality assurance/quality control and laboratory analysis

And this was all the Harvest that I reaped —
"I came like Water, and like Wind I will go."
Rubaiyat of Omar Khayyam
Edward Fitzgerald, 1809–1883

Quality assurance and quality control

Introduction

Programs that generate data of acceptable quality need to include a quality assurance (QA) component. Quality assurance encompasses management, procedures, and controls and includes a day-to-day quality control operational component.[17] Quality assurance is a set of operating principles that, if strictly followed during sample collection and analysis, will produce data of known and defensible quality.[1]

Quality control (QC) is a series of procedures or activities undertaken to ensure that the data meet appropriate standards. The primary goal of QC is to ensure that the sampling and analysis protocols are being properly executed and that errors are recognized before seriously impacting the data. *A QA plan provides the concept and theory, and a QC plan provides specific step-by-step instructions within the QA plan.*

Data quality objectives (DQOs) need to be identified during scoping of the project and the development of sampling and analysis plans.[13] Data quality objectives describe the overall level of uncertainty which the decisionmakers are willing to accept from environmental data and assist in specifying the objectives of a project, including developing and instituting acceptable QA/QC plans.

Quality assurance

A QA plan needs to be developed and implemented for the sampling of ground water monitoring wells and for the subsequent laboratory analysis. The QA plans describe the overall policies, procedures, standards, specifications, and documentation necessary to meet the DQOs and document that the specified QC activities are being followed; the documentation should be verifiable and defensible.[8]

The objectives of QA are to ensure that the data are meaningful, representative, complete, precise, accurate, comparable, and admissible as legal evidence.[12] A QA plan for ground water sampling specifies the details for calibration and maintenance of sampling equipment, the equipment and techniques involved in the sampling process, and the number and type of control samples.

A formal quality assurance project plan (QAPjP) is "an orderly assemblage of detailed procedures designed to produce data of sufficient quality to meet the DQOs for a specific data collection activity."[17] A QAPjP sets forth a plan for sampling and analysis which will generate data of a quality commensurate with its intended use. It includes QA/QC controls for both sample collection and laboratory analysis and is much more inclusive than a sampling plan. A QAPjP includes:[11]

1. A title page
2. A table of contents
3. A description of the project
4. Description of the organization and respective responsibilities
5. QA objectives
6. Sampling procedures
7. Sample custody
8. A description of the procedures used and the frequency of field and laboratory calibrations
9. Analytical procedures
10. Data reduction, validation, and reporting
11. Internal QC checks
12. Performance and system audits
13. Quality assurance reporting procedures
14. Preventative maintenance procedures and schedule
15. Specific standard operating procedures used to assess precision, accuracy, representativeness, comparability, and completeness (PARCC)
16. A description for correcting out-of-control situations

Some environmental programs require more formal and stringent QA/QC than other programs. The USEPA requires that a formal QAPjP be submitted to receive federal funding for environmental work undertaken for programs such as CERCLA and recommends them for RCRA. If a customer has little interest or requirement for a formal QA plan, then a consulting firm that routinely institutes a comprehensive QA/QC program may lose the contract to another consultant who provides less of a QA/QC effort.[18] It may appear that instituting adequate QA/QC is expensive. It is more likely to be much more expensive to have ground water data declared inadmissible (potentially in a court of law) because the QA/QC documentation was not of sufficient quality or quantity to defend it than it is to institute adequate QA/QC at the onset of a project.

The individual(s) responsible for a QA program should if possible, be entirely separate and independent from the personnel who are undertaking the work being evaluated. Those responsible for the management of a project must ensure that[17]

- Appropriate methodologies are followed as documented in the QAPjP
- Personnel clearly understand their duties and responsibilities
- Each staff member has access to the appropriate project documents
- Any deviations from the QAPjP are communicated to the project management
- Communication occurs between the field, a laboratory, and project management as specified in the QAPjP

All personnel involved in sampling, analysis, data reduction, and QA need to be sufficiently trained to ensure high-quality data. Appropriate training for personnel includes (1) on-the-job training, (2) a short-term (2 weeks or less) training course, and (3) a long-term (quarters or semesters) training course.[12] The QA plan should provide for a periodic assessment of training needs and describe the manner in which training is to be accomplished.

On-site evaluations should be a component of a QA plan and should be undertaken internally and externally. An on-site evaluation includes a review of the facilities, staff, training, instrumentation, procedures, documentation, methods, sample collection, QA policies and procedures, sample transport, and subsequent laboratory analysis. Such an evaluation needs to be carefully documented, and any problems which may cast aspersions on the data need to be immediately conveyed to the project manager.

Periodically, QA reports should be generated for ground water sampling and analysis and should include:

1. Changes in the approved QA plan
2. Significant problems, accomplishments, and status of corrective actions
3. Results of performance and systems audits
4. Assessment of the data in terms of precision, accuracy, representativeness, comparability, and completeness (PARCC)
5. Personnel training

Quality control

Essential elements of QC for ground water sampling include proper calibration of field measurement equipment, assurance of representative samples, and use of proper sample handling precautions.[7] The use of control samples helps to correct systematic errors introduced through collection, handling, storage, transport, and laboratory procedures. Optimizing PARCC is a goal of QC; PARCC is described as follows:

- *Precision* (reproducibility) is a measure of the probability that a measurement will fall within certain confidence limits. Precision is measured by repeated analysis of the same sample, and may be expressed graphically or in terms of standard deviations (e.g., 6.3 ± 1.3 mg/L).
- *Accuracy* (closeness to a "true" value) is measured indirectly by using reference materials or matrix spikes (such as percent recoveries) and is the degree of agreement between the measured value and the true value. Accuracy is an expression of the amount of bias in the data, and bias is defined as a systematic deviation (error) in the data. Accuracy may be assessed by the use of standard reference materials and is expressed as a percentage of the ratio of measured value to true value.
- *Representativeness* is the degree to which data accurately and precisely represent the ground water being sampled. Representative samples statistically are a subset consisting of the average characteristics of the set. Selecting the best sampling equipment and techniques assists in ensuring representative data. Using the field parameters and chemistry of the samples may also assist in determining the representativeness of ground water samples, as some combinations of concentrations and field parameters may not be possible.
- *Completeness* is the amount of valid data obtained for a measurement, compared to the expected amount of data required for the measurement. Completeness expresses the fraction of measurements that are not identified as "outlyers", using the following expression:

$$\text{Percent completeness} = \frac{\text{(number of valid data points)}}{\text{(total number of data points from collected samples}} \times 100$$

  Large deviations between sampling events (vs. a gradual trend) that could be attributed to sampling, analysis, or transcription errors should be flagged for review.
- *Comparability* is the ability to compare one data set, measuring system, or piece of equipment with another. Data should be comparable from site to site. The main reasons why data may not be comparable from site to site include the use of different sampling equipment, techniques, or even sampling crews; the sample handling was different (e.g., filtered vs. unfiltered); or different laboratory analytical procedures with differing limits of quantitation.

A QC plan specifies the number of control samples needed to reduce PARCC errors. Control samples are QC samples introduced into a process to monitor the performance of the system.[17] A good QC plan for a ground water sampling effort incorporates the following control samples, but not necessarily all of them, during the same sampling event:

- *Trip blanks:* A trip blank is a sample container filled by the analytical laboratory with reagent grade deionized water. Trip blanks are placed with the other clean (empty) sample containers and accompany the sample containers through the entire sampling and analytical process. Trip blanks are not opened by field personnel and are used to check for contamination in the sample bottles or in handling and/or shipment. At minimum, analysis of one trip blank per shipping container should be required. It may be a good practice to put two or three trip blanks in each shipping cooler, so the other blanks can be a analyzed and the results interpreted in case the first trip blank shows unacceptable contamination. Trip blanks are usually used only when collecting VOC samples.

- *Field Blanks:* Field blanks are prepared by pouring laboratory-supplied deionized water into clean sample containers in the field in the same location(s) where ground water samples are taken. Field blanks provide much of the same information as trip blanks, but also add information about potential sources of contamination derived from ambient conditions such as from automobile exhaust, dust, precipitation, etc. If field blanks are "clean" as per the laboratory analysis, then the trip blanks may not need analysis.
- *Duplicate samples:* Duplicate samples are samples taken from the same monitoring point, one immediately after the other, using the same equipment. Duplicate samples are a QC check on the laboratory methodology for ascertaining precision. One duplicate sample per day should be considered the minimum.
- *Rinsate (equipment)blanks:* Rinsate blanks are obtained by running deionized water through all the cleaned surfaces of the sampling equipment that the sample water contacts during sampling, including the insides of the pump and tubing, filters if used, ambient air, preservatives, and sample containers. These samples verify the efficacy of the equipment cleaning procedure and should be taken after cleaning the equipment and subsequent to sampling the most highly contaminated well in the network. The concentration of any contaminant found in a blank must not be used to correct ground water data.[17] At minimum, one rinsate blank per day should be required.
- *Split (laboratory) samples:* A split sample is similar to a duplicate sample, except that part of a sample is analyzed by one laboratory and the other part is analyzed by a separate laboratory. This is undertaken to evaluate a laboratory's ability to adequately analyze ground water samples (precision).
- *Matrix spiked samples:* Matrix spiked samples are samples to which a known quantity of a chemical constituent is added. They are then sent to the analytical laboratory with the other samples, without informing the laboratory that the samples were spiked. The spiked sample containers should look exactly like the other nonspiked samples when shipped to the analytical laboratory, including adding preservatives as appropriate and labeling the containers with fictitious but plausible well numbers. Uncontaminated upgradient ground water (the matrix) is spiked instead of deionized water since deionized water may be easily spotted by the laboratory. Spiking of samples is undertaken to

measure accurately the laboratory's ability to analyze ground water samples. The less known about the ability of an analytical laboratory, the more frequently spiked samples should be used. In general, 1 spiked sample per 20 ground water samples should be considered.

A large variety of control samples for addition to deionized water or for spiking ground water samples are available from several manufacturers. Inquiries can be directed to the USEPA Office of Research and Development, Environmental Monitoring Systems Laboratory, Cincinnati, OH 45268. These control samples include parameters that are routinely sampled in ground water (e.g., anions and cations, trace metals, and VOCs). The check samples come in sealed glass ampoules and are added for dilution to samples prior to analysis.[15]

Background (upgradient) samples are not strictly considered control samples but provide a standard against which ground water samples from downgradient monitoring wells are evaluated. At least one background sample should be required for each sampling event.

Data quality objectives

Data quality objectives (DQOs) are qualitative and quantitative statements that specify the quality of the data required to support decisions for remedial response activities.[13] They specify the amount of uncertainty or error allowed in the data and differ from precision and accuracy in that DQOs are limits for the overall uncertainty of the results, while the latter two are limits only for the uncertainty of specific measurements.

The DQOs specify thc quality of the data necessary to make decisions. Five analytical levels of data quality have been defined:[16]

1. *Field Screening (DQO Level 1):* This level provides the lowest data quality, but the most rapid results. This level is often used for initial site characterization, safety monitoring, preliminary comparison to standards, and engineering screening of bench-scale tests. Use of an organic vapor monitor (OVM), a pH meter, etc., are examples of equipment which provide DQO Level 1 data.
2. *Field Analysis (DQO Level 2):* This level provides rapid results and better-quality data than level 1. This level includes analyses from mobile laboratories, depending on the level of QA used.
3. *Engineering (DQO Level 3):* This is an intermediate level of data quality provided by some analytical laboratory methods which has a quick turnaround time but lacks full quality control documentation.
4. *Confirmational (DQO Level 4):* This level provides the highest data quality and is used for purposes of risk assessment, evaluation of remedial alternatives, and potential responsible party determination. These analyses require full Contract Laboratory Program (CLP) analytical and data validation procedures in accordance with USEPA-recognized protocol.

5. *Nonstandard (DQO Level 5):* This refers to analyses by nonstandard protocols such as for detecting unusual compounds or for analyzing compounds with very low concentrations. This type of analysis may require a method adaptation or a new method developed for the analysis. The level of QC is usually similar to DQO Level 4 data.

Qualitative DQOs include descriptions of actions that must be taken if the DQOs are not met; quantitative DQOs contain specific terms such as standard deviations, relative standard deviations, percent recovery, relative percent difference, and concentration.[9] Data quality objectives need to be identified, developed, and continually reevaluated for the entire process of site evaluation and remediation.

Laboratory analysis

The analytical laboratory

Analytical laboratories need to be constructed to facilitate proper analysis, have all the necessary equipment required for the specified test methods, have experienced staff familiar with the equipment and methods, follow the prepared standard operating procedures (SOPs) for the analysis of samples, and document fully the analytical procedures used. The function of the laboratory is to provide analytical results commensurate with the intended use, achieved through effective use of a QA program.[2] When samples are delivered to the laboratory, a set of documented procedures is followed for their receipt, sample handling, scheduling analysis, and sample storage. The SOPs need to specify routine laboratory practices such as how equipment is calibrated and maintained, how reagents, deionized water, and standards are prepared, how glassware is cleaned, pipetting techniques, etc.

The analytical work used to assess concentrations of constituents in ground water samples must be undertaken using USEPA-approved analytical methods whenever possible. If a USEPA method has not been developed for analyzing a particular sample (e.g., analyzing gasoline), then an adaptation of a USEPA method or use of other recognized methods such as those provided by ASTM (American Society for Testing and Materials) or APHA (American Public Health Association) may be necessary.

The test method used must be described exactly and even more so if not strictly undertaken as described in a published method. Documenting the actual laboratory procedures should include the following:

1. Sample preparation and analysis procedure
2. Instrument standardization
3. Sample data
4. Precision and bias
5. Detection and reporting limits
6. Test-specific QC

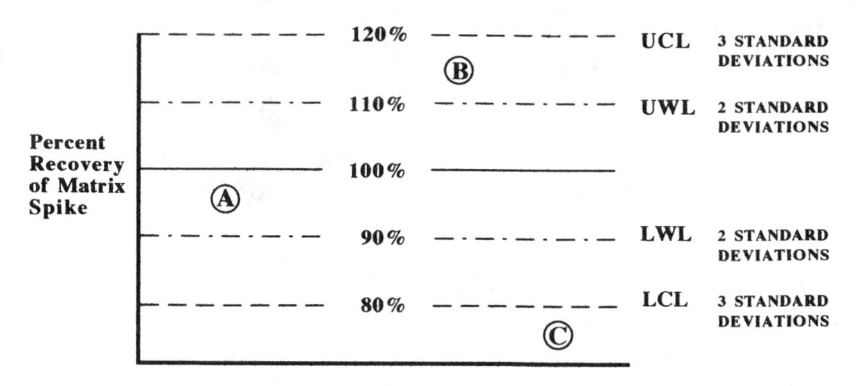

Figure 7.1 A typical analytical control chart.

The steps that identify and correct analytical deficiencies need to be specified. These steps include preparing new standards, recalibrating or restandardizing equipment, reanalysis of samples, or additional training of laboratory personnel in methods and procedures.[17] Quality assessment uses internal and external QC measures to determine the quality of the data produced by the laboratory. A good laboratory QC program incorporates recovery of known additions, analysis of externally supplied standards, analysis of reagent blanks, calibration of standards, analysis of duplicates, and maintenance of control charts.[1]

Procedures need to be in place for determining method proficiency. These include determining the precision and bias and the method detection limits (MDLs); control limits must be established that are limits of precision and bias. Typically, precision and bias are determined by analyzing duplicate samples or by adding a compound to the sample (a matrix spike) not usually found in such samples. The percent recovery of the matrix spike is used to assess the analytical ability of the laboratory and procedures used.

A compound commonly used as a matrix spike for assessing adequate recovery of VOCs is bromofluorobenzene. For example, if a known quantity of bromofluorobenzene is added to the sample and less than 70% is recovered, then the analysis may be considered inadequate. Figure 7.1 is a typical control chart for assessing the quality of the data.

The acceptance limits for recoveries of analytical parameters may vary from program to program, depending on the DQOs. Typical acceptance limits for duplicate samples (or matrix spikes) of ground water samples are as follows:[1]

Metals:	80-120%
VOCs:	70-130%
Anions:	80-120%
Nutrients:	70-130%
Herbicides:	40-160%

Analytical laboratories should never try to correct analysis results for loss of the matrix spike or for differences between duplicate samples, but need to report exactly what was found and the precision of the analysis. It may be common for a laboratory to run a daily check standard of all the compounds for analysis, instead of recalibrating the equipment to verify that they are within certain limits (e.g., ±10%). Some of the compounds may be out of the prescribed limits using the daily check standard, but most of them need to be within the prescribed limits. If the equipment is out of limits for certain analytes, the data need to be qualified as such in the report.

A procedure for reviewing and validating analytical data needs to be documented. This is undertaken by interpreting the results of QC samples and other independent procedures used to ensure that the analyses have been run correctly. Errors, deficiencies, deviations, or laboratory events or data that fall outside of established acceptance criteria need to be investigated, corrected, and documented.

Preparing samples for analysis

The analytical procedure selected must take into account the concentration of the sample. Sample concentrations have been defined as: trace (<1000 ppm), minor (1000 to 10,000 ppm), and major (>10,000 ppm). Many of the USEPA methods for drinking water are run using trace concentration analyses. Preparation techniques that optimize sample concentration for analysis include:

- Adjustment of the size of the sample prepared for analysis
- Adjustment of injection volumes
- Dilution or concentration of the sample
- Elimination of concentration steps prescribed for "trace" analyses
- Direct injection of samples to be analyzed for volatile constituents

Methods for preparing substances to be analyzed in ground water samples include the purge-and-trap method for volatile organic compounds, extraction by solvents for semivolatile compounds, and the use of acid extraction followed by filtration for metals analyses.

The purge-and-trap method for concentrating VOCs is undertaken by sparging (bubbling) 5 mL of the ground water sample with an inert gas in a water column at least 3 cm deep and collecting the volatilized VOCs on an adsorbent (called the trap) (Figure 7.2). When purging is complete, the sorbent tube is heated and backflushed with helium to desorb the trapped sample components into the packed gas chromatography column.[14]

A solvent such as methylene chloride or hexane may be used for dissolving semivolatile compounds, and after further preparation the extract is injected into a gas chromatography column equipped with a specialized detector(s). Preparing samples for metals analysis includes digesting the sample by adding dilute or concentrated nitric or hydrochloric acid, followed by heating the sample and reducing its volume, followed by filtration of the digestate. The prepararation of ground water samples for metals analysis depends on the method chosen to analyze the samples and for what the sample is to be analyzed (total metals, dissolved metals, etc.).

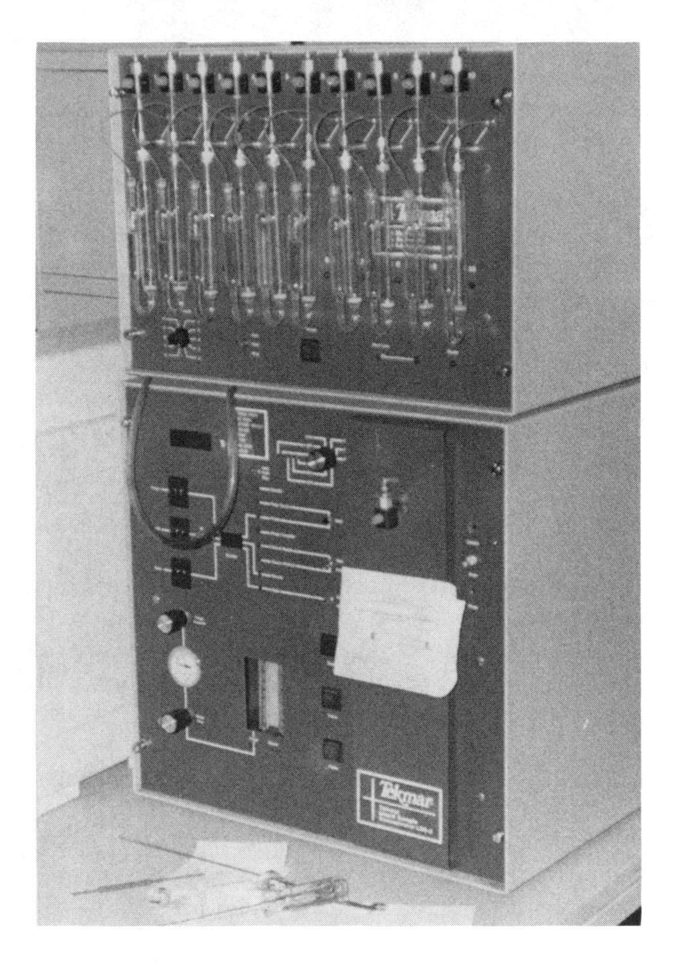

Figure 7.2 Purge-and-trap equipment.

Purification of samples (referred to as sample cleanup) may be required to remove potential interferents or substances which can distort peaks or shorten the life of expensive columns. Cleanup techniques include partitioning between immiscible solvents, adsorption chromatography, gel permeation chromatography, distillation, and chemical destruction of the interfering substances with acid, alkali, or oxidizing agents. Combinations of these cleanup techniques may also be required.

Selecting an appropriate analytical method

A dichotomy between detection samples and monitoring samples may be made for analytical purposes. Detection of organic substances in samples may be best accomplished by use of a mass spectrometer, which is more specific but less sensitive than using gas chromatography and which may lead to fewer false positives. Monitoring samples are used to confirm ongoing conditions, i.e., tracking the presence or absence of constituents in an environmental matrix such as ground water.

Analytes are divided into classes based on the determinative methods used to identify and quantify them. Tables 2-2 through 2-20 in Volume 1A of SW-846[17] list the acceptable USEPA analytical methods for analyzing classes of compounds, including (but not limited to) the following:

- Base/neutral fraction
- Acid fraction
- Halogenated volatiles
- Nonhalogenated volatiles
- Aromatic volatiles
- Phenols
- Organochlorine pesticides and PCBs
- Nitroaromatics and cyclic ketones
- Haloethers
- Chlorinated hydrocarbons
- Organophosphorus compounds
- Chlorinated herbicides
- Volatiles
- Semivolatiles
- Dioxins and dibenzofurans
- Polynuclear aromatic hydrocarbons
- Inorganic elements and compounds

The method chosen for analysis of a particular sample is highly dependent on the objectives of the analysis. Several analytical methods may be appropriate for particular constituents, and there is some overlap in the analytical techniques used for the groupings listed above. The detection limits of an analytical procedure may need to be based on the National Primary Drinking Water Standards. For example, the procedure for nitrate-N needs to be able to detect concentrations well below 10 mg/L, which is the drinking water standard for nitrate-N.

Numerous methodologies for analysis of various environmental samples have been compiled and published. These are periodically updated and refined, including:

1. USEPA Method 500: Analysis of drinking water for various contaminants
2. USEPA Method 600: Various methods for water and waste water
3. USEPA SW-846: Analysis of RCRA environmental samples
4. ASTM Methods: Various test methods for different industries
5. Standard Methods for Analysis: Varying matrices and analyses
6. Standard Methods for the Examination of Water and Wastewater (APHA): Water analyses for various analytes

Methods 1, 3, and 5 are most commonly used for environmental analyses.

Methods for determining organic compounds in ground water

The three determinative methods for organic compound analysis in ground water samples are gas chromatography (GC), high performance liquid chromatography (HPLC), and gas chromatography/mass spectrometry (GC/MS).

Gas chromatography

Gas chromatography encompasses all chromatographic methods in which the moving phase is a gas.[3] Columns may be long (up to tens of meters), have small diameters, consist of straight or coiled tubing (usually made of stainless steel or fused silica), and contain a variety of polar and nonpolar stationary phases. Stationary phases consist of a dry granular solid, a liquid supported by the granules or the walls of the column, or both. Components of a sample are advected through the column with an inert, high-purity carrier gas (typically helium) at different rates and times due to differential adsorption of the vapor phase onto the stationary phase. Gas-liquid chromatography is essentially the same as gas chromatography, except the stationary phase is a thin film of a nonvolatile liquid coated on a finely divided inert support with a large surface area. Under the Certified Laboratory Program (CLP) either one column and two detectors or two columns and one detector are required for analysis. One type of detector (the electrolytic conductivity, or Hall detector) is required to "see" single-bonded halogenated compounds, and another detector (photoionization) is required to "see" double-bonded nonhalogenated compounds. The use of two columns thus only doubly confirms the presence of double-bonded halogenated species. If a second column is utilized (such as one with a different polarity), it is placed in series downstream from the first column to confirm GC identifications, or resolve coeluting compounds.

The previously prepared sample (diluted, concentrated, etc.) is injected with a syringe into the sample port, where it is vaporized and swept along with the flow-controlled carrier gas into and through the column. Effluent from the column passes to a detector, where the components are measured and recorded. A detector (flame ionization, electron capture, flame photometric) indicates the presence of eluted components and transforms them into electrical signals, which are then represented as chromatograms. A chromatogram is a graphical representation of the detector response vs. time, or effluent volume (Figure 7.3). The time it takes for a peak to occur on the chromatogram is related to the molecular mass, boiling point, and chemical composition or polarity of the compound. The amplitude or area under the peak represents the quantity of the compound. The chromatogram is compared to standards either visually or with a computer (against an electronic library of standards) to determine what compound(s) individual peaks represent. The amount of separation in a column depends on differences in the distribution of volatile compounds (organic or inorganic) between a gaseous mobile phase and the selected stationary phase in the column.[6] Temperature control of the column is important for comparing elution rates, and analytical

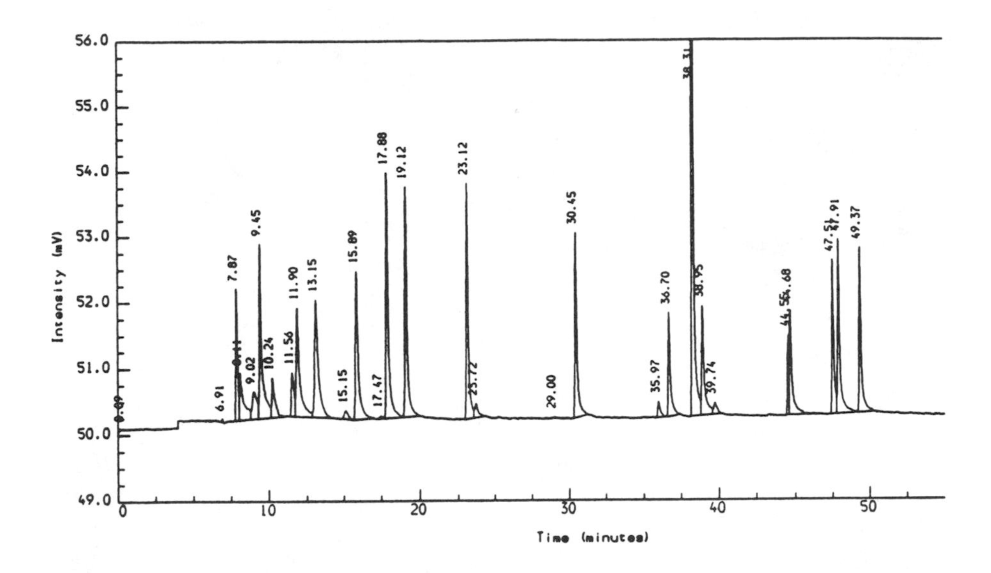

PEAK INFORMATION

RT mins	RT Exp	Hght uV	ug/L	Peak name
0.089	0.000	27	0.00030	
6.911	0.000	51	0.00056	
7.871	7.910	1996	1.83271	DICHLORODIFLUOROMETHANE
8.111	0.000	718	0.00798	
9.022	9.060	419	2.08906	CHLOROMETHANE
9.449	9.450	2650	1.56105	VINYL CHLORIDE
10.244	0.000	599	0.00665	
11.556	11.570	666	1.96304	BROMOMETHANE
11.898	11.880	1645	2.40977	CHLOROETHANE
13.147	13.140	1780	1.56812	TRICHLOROFLUOROMETHANE
15.147	0.000	130	0.00144	
15.889	15.870	2256	1.37118	1,1-DICHLOROETHENE
17.471	17.430	48	1.82065	ALLYL CHLORIDE
17.880	17.870	3733	1.29536	METHYLENE CHLORIDE
19.124	19.090	3497	1.37041	trans-1,2-DICHLOROETHENE
23.124	23.070	3581	1.33304	cis-1,2-DICHLOROETHENE
23.720	23.630	215	0.06611	CHLOROFORM
29.000	28.870	20	0.00742	TRICHLOROETHENE
30.449	30.360	2814	1.21355	BROMODICHLOROMETHANE
35.969	35.850	242	0.04209	TETRACHLOROETHENE
36.702	36.630	1570	1.20823	CHLORODIBROMOMETHANE
38.307	38.260	12783	12782.74512	4-FLUOROCHLOROBENZENE
38.947	38.930	1651	1.30780	CHLOROBENZENE
39.738	0.000	174	0.00193	
44.551	44.480	1213	1.30150	2-CHLOROTOLUENE
44.684	44.630	1581	1.35642	4-CHLOROTOLUENE
47.507	47.450	2338	1.30540	1,3-DICHLOROBENZENE
47.911	47.840	2643	1.39996	1,4-DICHLOROBENZENE
49.369	49.310	2504	1.31609	1,2-DICHLOROBENZENE

Figure 7.3 A typical chromatogram for volatile organic compounds.

equipment can be programmed to heat or cool the column in steps to assist in eluting the sample.

High performance liquid chromatography

High performance liquid chromatography typically uses shorter columns than those used with GC analysis, and they are packed with a variety of

Figure 7.4 High performance liquid chromatography equipment.

materials which comprise the stationary phase; HPLC uses a liquid (such as high-purity water) as a carrier instead of a gas. If water is used as the carrier, then the pH of the water can be adjusted to change the elution on the stationary phase. Ultraviolet detectors are commonly used with HPLCs, and the HPLC may be used for analysis of compounds such as phenols, polycyclic aromatic hydrocarbons, and carbamates (Figure 7.4).

Mass spectrometry

A mass spectrometer detects compounds based on retention time and determines what type of compound is present based on a mass spectral breakdown. The gaseous sample is introduced into a MS through a large magnetic or electric field which shatters the compound into fragments (positive ions).

Separation of the beam of ions according their respective masses is usually undertaken by magnetic deflection or a quadrapole filter. Other separating techniques include time-of-flight, radio frequency, cyclotron resonance, and cycloidal focusing. The voltage required to ionize molecules is approximately 10 eV, and a typical MS uses 70 eV. The mass spectrometer must be capable of scanning from 35 to 260 amu every 3 seconds. The detector matches up the mass spectra thus created with known spectra, as individual compounds always break down in the same way. Continuous changing of the magnetic or electric field that effects separation produces a mass spectrum. A mass spectrum depicts the mass of the molecule and the masses of the pieces from it.[10]

The three principal means of identifying compounds from mass spectra are manually comparing mass/intensity data, computerized spectral matching, and interpretation based on an understanding of the process of fragmentation of organic molecules.

Gas chromatography/mass spectrometry method

The GC/MS method has the advantage of using two relatively independent means of determining volatile constituents in the matrix. The sample is introduced into the GC column, is eluted, and is then passed into the MS. Scanning may be triggered manually or scanned continuously under the control of a computer, and the MS scan numbers are directly correlated to retention times from the GC.

Two criteria must be satisfied in order to compare a sample to a standard using the GC/MS method:[3]

1. Elution of the sample component at the same retention time as the standard component, as shown by co-injection or standard addition
2. Correspondence of the mass spectrums of the sample and standard components.

In order to confirm the presence of a compound, both the GC retention data and the mass spectrum of a compound must uniquely match those of a co-injected authentic standard reference compound.

Methods for determining metals in ground water

The methods most commonly used for metal analyses of ground water samples include:

- Inductively coupled plasma argon emission spectroscopy (ICP-AES)
- Inductively coupled plasma mass spectroscopy (ICP-MS)
- Flame (direct-aspiration) atomic absorption spectrometry (FLAA)
- Graphite furnace atomic absorption (GFAA)

Sensitivity for detection of metals in ground water samples is generally ICP < FLAA < GFAA.

The ICP-AES method utilizes high-resolution optical spectrometry to measure element-emitted light. The prepared liquid sample is reduced to a fine spray (nebulized), and the resultant aerosol is transported to the high-temperature ionized-gas (plasma) torch. Atomic-line emission spectra that are element-specific are produced by a radio-frequency inductively coupled plasma. The spectra are dispersed by a grating spectrometer, and the intensities of the lines are monitored by photomultiplier tubes. The method specifies a recommended wavelength for each chemical element.

The ICP-MS method measures ions produced by a radio-frequency inductively coupled plasma. The sample is nebulized, and aerosol is transported by argon gas into the plasma torch. The ions produced are trapped in the plasma gas and are introduced, by a water-cooled interface, into a quadrupole mass spectrometer.[17]

The FLAA method uses either a nitrous oxide/acetylene or air/acetylene flame as an energy source for dissociating (atomizing) the aspirated sample into the free atomic state, making analyte atoms available for absorption of light. A hollow cathode lamp or a discharge lamp without electrodes provides a beam of light that passes through the flame into a monochromator (which isolates the characteristic radiation from the lamp) and onto a detector that measures the amount of absorbed light. The wavelength of the light beam is chosen such that it is specific to the metal being analyzed, so the amount of light absorbed is a measure of the concentration of the metal in the sample.

Operating principles of the GFAA method are similar to those used with FLAA, but the GFAA uses an electrically heated graphite furnace instead of a flame, which allows heating of the sample aliquot in several stages to remove unwanted matrix components such as organic and inorganic molecules and salts. The graphite tubes used must be compatible with the furnace; a conservative estimate is that a tube will last at least 50 firings.[3] The GFAA method has extremely low detection limits but is very sensitive to interferences (Figure 7.5).

Data management and the analytical report

A procedure for managing records is part of laboratory operations, including the specifics of record generation and control and the requirements for the authority to govern record retention (including type, time, security, retrieval, and disposal). Data from analysis of samples needs to be reduced, reviewed by a second analyst or supervisor, and presented in a way (preferably electronically) to facilitate decisions about the site where the sample was obtained. The following information is recommended for reporting results from the analytical laboratory.

The report cover sheet

1. Name, address, and phone number of the analytical laboratory
2. Names of the sampler(s) and analyst(s) (and signatures, as appropriate)
3. Name of the site where the samples were obtained and all the stations that were sampled (i.e., MW-1 and MW-7)
4. Date and time the samples were collected
5. The date the report was prepared
6. Analytical test methods used and any significant deviations from the approved method(s)
7. Method detection limits (MDLs) or reporting limits that reflect dilutions, interferences, or corrections for dry weight

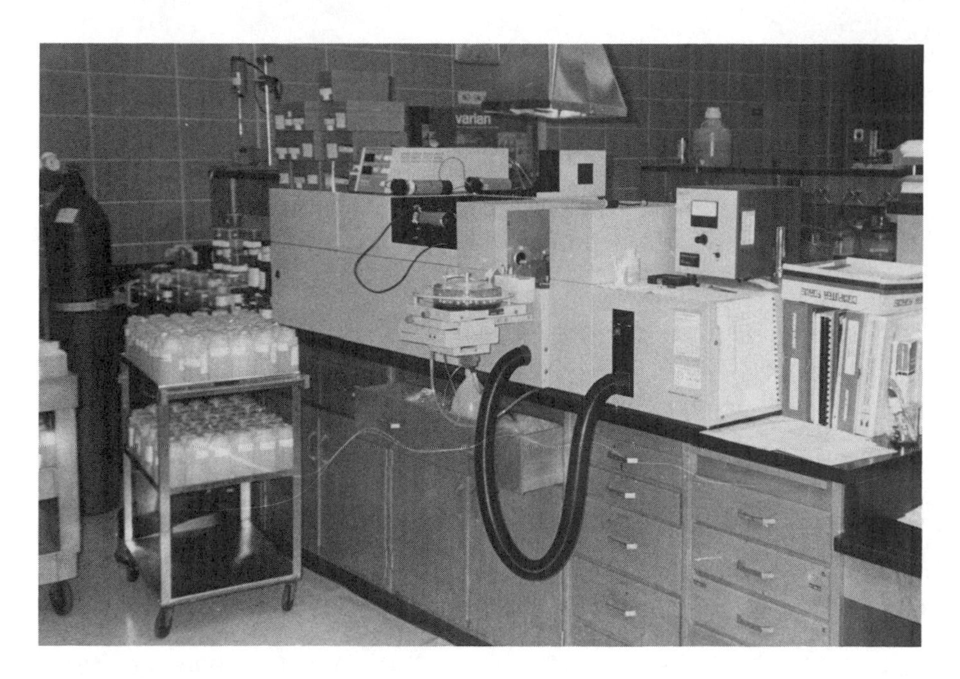

Figure 7.5 Graphite furnace atomic absorption equipment for analyzing metals in ground water.

The analysis sheet

1. Name of the site
2. Sampling station (e.g., MW-5)
3. Date analyzed (date of extraction for organic compounds)
4. Laboratory I.D. number
5. Analyte by grouping (e.g., purgable aromatic)
6. Data in columns under the following headings: **Analyte, MDL, Standard/MCL, Results** (for regulatory purposes, each exceedance of the standard/MCL should be highlighted or underlined)
7. A (brief) narrative on how the analysis matched the analytical DQOs. This should include appropriate QC results (correlation with the sample batch should be traceable and documented), data qualifiers with appropriate references, and a narrative on the quality of the results

Appendices

1. Photocopies of the original field parameter and stabilization sheets and field notes
2. Narrative on any departures from standard field protocols
3. Chain-of-custody records

Upon receipt of the analytical report, a review of the laboratory analysis of the check samples and all data qualifiers (field and laboratory) needs to be undertaken by the individuals who have charge over the site. Depending on the regulatory status of the site, decisions are then made on appropriate actions required if violations of drinking water or surface water standards occurred. If a performance standard (intervention limit or MCL) was exceeded in a monitoring point, at least one more sample from the same well (a confirmational sample) may be needed and, depending on the specifics of the site, the scope of the investigative effort expanded.

References

1. APHA, *Standard Methods for the Examination of Water and Wastewater*, 18th ed., American Public Health Association, Washington, D.C., 1992.
2. ASTM, *Standard Practice for Good Laboratory Practices in Laboratories Engaged in Sampling and Analysis of Water*, ASTM D 3856-88, American Society for Testing and Materials, Philadelphia, PA, 1988.
3. ASTM, *Standard Practice for Gas Chromatography Terms and Relationships*, ASTM E 355-77, American Society for Testing and Materials, Philadelphia, PA, 1989.
4. ASTM, *Standard Practice for Identification of Organic Compounds in Water by Combined Gas Chromatography and Electron Impact Mass Spectrometry*, ASTM D 4128-89, American Society for Testing and Materials, Philadelphia, PA, 1989.
5. ASTM, *Standard Practice for Measuring Trace Elements in Water by Graphite Furnace Atomic Absorption Spectrophotometry*, ASTM D 3919-85, American Society for Testing and Materials, Philadelphia, PA, 1989.
6. ASTM, *Standard Practice for Packed Gas Chromatography*, ASTM E 260-91, American Society for Testing and Materials, Philadelphia, PA, 1991.
7. Barcelona, M.J., J.P. Gibb, J.A. Helfrich, and E.E. Garske, Practical Guide to Groundwater Sampling, Illinois State Geological Survey Contract Report 374, Springfield, 1986.
8. Dillaha, T.A., S. Mostaghimi, C.D. Eddelton, and P.W. McClellan, *Quality Control/Quality Assurance for Water Quality Monitoring*, American Society of Agricultural Engineers, St. Joseph, MO, 1988.
9. Keith, L.H., *Environmental Sampling and Analysis: A Practical Guide*, Lewis Publishers, Chelsea, MI, 1991.
10. McLafferty, F.W., *Interpretation of Mass Spectra*, University Science Books, Mill Valley, CA, 1980.
11. USEPA, Guidelines and Specification for Implementing Quality Assurance Requirements for EPA Contracts and Interagency Agreements Involving Environmental Measurements, QAMS-002/80. Office of Research and Development, U.S. Environmental Protection Agency, Washington, D.C., 1980.
12. USEPA, Handbook for Sampling and Sample Preservation of Water and Wastewater, EPAA-600 4-82-029, U.S. Environmental Protection Agency, Washington, D.C., 1982.
13. USEPA, Data Quality Objectives for Remedial Response Activities, EPA/540/G-87/003, U.S. Environmental Protection Agency, Washington, D.C., 1987.
14. USEPA, Methods for the Determination of Organic Compounds in Drinking Water, EPA/600/4-88/039 (500 Series for SDWA), U.S. Environmental Protection Agency, Washington, D.C., 1988.
15. USEPA, Newsletter Quality Assurance, Vol. 11 No. 2, U.S. Environmental Protection Agency, Washington, D.C., 1990.

16. USEPA, Region V Model Superfund Quality Assurance Project Plan (QAPjP), U.S. Environmental Protection Agency, Washington, D.C., 1991..
17. USEPA, Test Methods for Evaluating Solid Waste, SW-846, U.S. Environmental Protection Agency, Washington, D.C., 1992.
18. Van EE, J.J., and L.G. McMillion, Quality assurance guidelines for ground water investigations: the requirements, *Ground-Water Contamination: Field Methods,* ASTM Spec. Tech. Publ. 963, American Society for Testing and Materials, Philadelphia, PA, 1986.

Index

E

F

G

H

I

J

L

M

O

P

Q

T

U

V

W